Passerine Migration

Nikita Chernetsov

Passerine Migration

Stopovers and Flight

Springer

Nikita Chernetsov
Biological Station Rybachy
Zoological Institute
Universitetskaya nab. 1
St. Petersburg 199034
Russia

ISBN 978-3-642-29019-0 e-ISBN 978-3-642-29020-6
DOI 10.1007/978-3-642-29020-6
Springer Heidelberg New York Dordrecht London

Library of Congress Control Number: 2012934250

Translation from the Russian language edition: Passerine Migration: Stopovers and Flight by Nikita Chernetsov, © KMK Scientific Press Ltd, Moscow 2010. ISBN 978-5-87317-698-4.

Printed on acid-free paper

Springer is part of Springer Science+Business Media (www.springer.com)

Preface

Even though the importance of migratory stopovers and the related events for bird migration as a whole has been intuitively clear since long ago, the targeted study of this field did not start until a relatively recent time. Since the late 1980s—early 1990s, stopover ecology and behaviour have attracted the attention of a number of students of bird migration. An important point was the dissertation of Åke Lindström 'Stopover ecology of migrating birds' presented at Lund University (Lindström 1990). At the same time Thomas Alerstam and Åke Lindström published their seminal paper that laid the foundation of the optimal bird migration theory (Alerstam and Lindström 1990). Equally important were the studies of Frank R. Moore and his colleagues at the University of Southern Mississippi (Moore and Kerlinger 1987; Loria and Moore 1990; Moore and Wang 1991) and the dissertation of Andreas Kaiser which was based on work done at what is now Max Planck Institute for Ornithology (Vogelwarte Radolfzell) and presented at Konstanz University (Kaiser 1993).

Even though the study of stopover ecology and behaviour of migrating birds actively developed and for some time was a 'hot topic' in bird migration research, rather surprisingly, no monographic review on stopover biology of birds has been produced. This book is an attempt to fill this gap.

My own research of stopover ecology and behaviour started when I, as an undergraduate student, was doing my diploma project on migration ecology of *Acrocephalus* warblers at the Biological Station Rybachy of the Zoological Institute of the Russian Academy of Sciences under the supervision of Casimir V. Bolshakov. Since 1994, I have worked at Rybachy doing research of stopover biology of songbirds, mainly long-distance Palaearctic-African migrants. Most of my fieldwork was done on the Courish Spit in Kaliningrad Region of Russia, which has a long tradition of avian migration research going back to the early 20th century when it was in the German province of East Prussia. The study of bird migration was one of the main fields of activity of the world's first bird observatory Vogelwarte Rossitten in 1901–1944 (Thienemann 1931; Vaughan 2009), and it has also been a very important working field of the Biological Station Rybachy which replaced it in 1956.

Some fieldwork was done at the Dzhabybek research station of the Institute of Forestry of the Russian Academy of Sciences (West Kazakhstan Region, Kazakhstan). I also analysed data collected at the International Birding and Research Center Eilat in Israel.

I am most grateful to the director of the Biological Station Rybachy of the Zoological Institute Casimir V. Bolshakov who was my supervisor and has always supported my work. I am obliged to the members of the Biological Station Rybachy Dina S. Lyuleeva, Vladimir A. Payevsky, Mark E. Shumakov, Natalia V. Vinogradova, Valentina P. Dyachenko (†), Vladislav D. Efremov, Leonid V. Sokolov, Anatoly P. Shapoval, Mikhail Y. Markovets, Alexandra Sinelschikova, Nadezhda P. Zelenova, Dmitry Leoke for their invariably kind attitude and constant help. The importance of the collaborative work and (often heated) discussions with Nikolay Titov, Andrey Mukhin, Vladislav Kosarev, Arseny Tsvey, Pavel Ktitorov, Dmitry Kishkinev and Vitaly Grinkevich cannot be overestimated.

Marina L. Sizemskaya and Mamay K. Sapanov made the work at the Dzhanybek field station possible and convenient. Reuven Yosef made the data from Eilat available for analysis. I am most grateful to Peter Berthold, Eberhard Gwinner (†), Franz Bairlein and Frank R. Moore for the possibility to visit their respective institutes and for fruitful discussions. Many discussions with the colleagues from various countries, especially with Julia Delingat, Michael Schaub, Volker Salewski, and Goetz Eichhorn were very constructive. The author is also most grateful for the comments made by the reviewers that made it possible to improve the drafts: Vladimir B. Zimin, Dmitry O. Eliseev, Vladimir A. Payevsky, Valery M. Gavrilov, Alexander V. Artemyev, Alexander V. Andreev. Tatiana A. Ganf provided linguistic help which is much appreciated. Ekaterina A. Papchinkaya helped to format the illustrations.

The author expresses a sincere gratitude to all these people. The responsibility for all the shortcomings remains with the author alone.

References

Alerstam T, Lindström Å (1990) Optimal bird migration: the relative importance of time, energy, and safety. In: Gwinner E (ed) Bird migration. Springer, Heidelberg

Kaiser A (1993) Rast- und Durchzugstrategien mitteleuropäischer Singvögel. Analysen von Fang- und Wiederfangdaten von Fangstationen zur Beschreibung der Ökophysiologie und des Verhaltens rastender Populationen. Doktorarbeit, Universität Konstanz

Lindström Å (1990) Stopover ecology of migrating birds. Dissertation, Lund University

Loria DE, Moore FR (1990) Energy demands of migration on red-eyed vireos, *Vireo olivaceus*. Behav Ecol 1:24–35

Moore FR, Kerlinger P (1987) Stopover and fat deposition by North American wood-warblers (Parulinae) following spring migration over the Gulf of Mexico. Oecologia 74:47–54

Moore FR, Wang Y (1991) Evidence of food-based competition among passerine migrants during stopover. Behav Ecol Sociobiol 28:85–90

Thienemann J (1931) Vom Vogelzuge in Rossitten. Neumann-Neudamm, Neudamm

Vaughan R (2009) Wings and rings: a history of bird migration studies in Europe. Isabelline Books, Falmouth

Contents

Chapter 1
Introduction

Abstract This chapter gives a general introduction to the migratory movements of birds, and outlines the ultimate factors that govern the migratory behaviour. The remaining part of this monograph mainly deals with the proximate controlling factors. It is emphasized that the realised behaviour of avian migrants results from an interplay of their endogenous migratory programme and the environmental conditions the migrants encounter en route. The definitions of avian migration, migratory stopover, its duration, fuel deposition rate, and departure fuel load accepted in this monograph are given in this chapter.

1.1 General Introduction to Migratory Movements of Birds

Many songbirds perform long-distance migratory movements twice a year. Annual movements of some passerines are no less impressive than the famous migration of Alaskan Arctic terns *Sterna paradisaea* to their winter quarters in Antarctica (Berthold 2001; Newton 2008; Egevang et al. 2010). The longest migratory movements among songbirds are probably performed by northern wheatears *Oenanthe oenanthe* that travel between Alaska and northeast Asia and East Africa, and by willow warblers *Phylloscopus trochilus* that breed in northeast Asia and winter in Central and East Africa. If migratory distance is taken relative to body size, this willow warbler population may be performing the longest migratory movements in the world (Alerstam 2003).

Not only such extreme long-distance migrants, but most other migratory birds do not cover the distance between their breeding areas and winter quarters by a single flight. Some very impressive exceptions do exist, e.g. bar-tailed godwits *Limosa lapponica baueri* that fly non-stop from Alaska to New Zealand across the Pacific (Gill et al. 2005, 2009). However, a vast majority of migrants make multiple stopovers *en route* during which they rest and refuel.

N. Chernetsov, *Passerine Migration*, DOI: 10.1007/978-3-642-29020-6_1,

In daytime migrants and in migrants with the mixed rhythm of migratory activity (Bolshakov 1977) it may not be easy to distinguish between migratory flights and stopovers. This is true of birds like titmice that make migratory movements by flying low above the canopy, so that it may be difficult to say whether they are migrating or foraging. Unlike many other songbirds, swallows and martins can fly and forage (Lyuleeva 1970), and American crows *Corvus brachyrhynchos* have been recently shown to do the same (Ward and Raim 2011). Detailed observations of the behaviour of moving passerines make it possible to tell foraging movements from diurnal migration, but often it is not easy to make the distinction.

In nocturnal migrants flight and stopovers are usually better separated. When migrating over areas with continuous stopover possibility, i.e. when not crossing ecological barriers, these birds perform long-distance flights (covering dozens and hundreds of kilometres) during the night and rest and forage within a much smaller area during the daytime. This diel pattern of flights and stopovers makes a clear distinction between them possible (Bairlein 1992). The linear size of a stopover area may be as long as dozens and hundreds of metres, at most several kilometres (Wang and Moore 1993; Titov 1999; Chernetsov et al. 2004; Chernetsov and Mukhin 2006; Paxton et al. 2008), and is anyway negligible as compared with the distance of single migratory flights which may be dozens and more often hundreds of kilometres (Bulyuk and Chernetsov 2000; Barriocanal et al. 2002; Hall-Karlsson and Fransson 2008). As shown by our studies on the Courish Spit in the Eastern Baltic, 10.7 km is at the very upper limit of the distance of daytime movements in songbird nocturnal migrants (Chernetsov 2011).

Only when normally nocturnal migrants are crossing large water bodies or other areas with very poor stopover opportunities (e.g. the Sahara), their diel rhythm of flight activity changes so that they can fly both during night and day. It happens when crossing the Gulf of Mexico (Gauthreaux 1971, 1972), the western Atlantic between North and South America (Williams and Williams 1990), the Mediterranean and the Baltic in their widest parts (Bruderer 2001), and under some situations (e.g. in spring) when crossing the Sahara (Schmaljohann et al. 2007). Therefore, nocturnal migrants are a suitable system for studying flights and stopovers when they migrate over good habitats with continuous stopover opportunities. Most data collected, analysed and discussed in this monograph refer to passerine nocturnal migrants.

It seems intuitively obvious that avian migration first of all is flight. However, careful analysis shows that the role of stopovers is very important in terms of both time and energy. Actual flight represents 29% of energy expenditure during migration, whereas the rest is spent on supporting the animals during stopovers (Wikelski et al. 2003). This proportion (roughly one-third during flight, two-thirds during stopovers) was predicted by the optimal migration theory (Alerstam and Hedenström 1998) and supported by direct measurements of flight costs during actual free migratory flight in *Catharus* thrushes by injecting doubly labelled water D_2O^{18} (Wikelski et al. 2003). However, it should be mentioned that recent tracking data on red-backed shrikes *Lanius collurio* during their migration between Europe and Africa showed a very high rate of movement, corresponding to travelling during 7–10 h daily, and even 11–16 h daily when crossing the Sahara (Tøttrup et al. 2012). Thus, some species may spend a large proportion of their time actually flying, but they apparently need quite some time to fuel up before such flight bouts, and this fuelling time should be included in the estimates.

Because of the important role that stopovers play in the temporal and energetic balance of migration, their significance for organisation of migration is paramount. Stopovers are mainly made for refuelling, not for rest (Dolnik 1985), when continuing flight is not optimal due to adverse environmental conditions, e.g. high air temperature causing negative water balance, headwind, low visibility, precipitation etc. (Dolnik 1990; Biebach 1990; Biebach et al. 1991, 2000; Schmaljohann et al. 2007). Thus, some stopovers are used for refuelling and some for avoiding adverse weather, with varying behaviour.

Probably it is during the stopover that orientation tasks are solved by the migrants. More and more data becomes available suggesting that calibration of compass systems used by migrants (sunset cues, geomagnetic field and possibly stellar cues) occurs at stopovers before take-off (Cochran et al. 2004; Muheim et al. 2006, 2007, 2009; but see Rabøl 2010; Chernetsov et al. 2011). During migratory flights migrants seem to be maintaining a selected direction of movement, more or less successfully due to wind drift (Moore 1990; Thorup et al. 2007).

Unlike many waders and waterfowl, songbirds can stop over at any site along their flyway, and are not restricted to the scattered patches of stopover habitats which are often traditionally used by many generations of migrants. This is called a continuous stopover opportunity as opposed to a discrete stopover opportunity (Houston 1998). This is certainly only true if songbirds are migrating over optimal or at least suitable habitats. If they are crossing an ecological barrier (e.g. sea or desert), they can only stop at few suitable sites (islands in the sea or oases in the desert). Under such conditions passerines are also facing a discrete stopover opportunity.

Stopover ecology and behaviour is a complex of ecological, physiological and behavioural adaptations aimed at solving the specific problems that migrants are facing. It is not identical to the so-called 'migratory syndrome' that does not seem to exist, in the sense that there is no evidence for deeply rooted co-adapted trait complexes (Piersma et al. 2005). However, there are some common challenges faced by avian migrants, e.g. they all should try to refuel as fast and efficiently as possible, and they all should try to survive. The migratory stopover may be a rare situation when the main assumption of the optimal foraging theory [i.e. that the animal's behaviour is aimed at maximising the net energy intake: Krebs and Davies (1991), Begon et al. (2006)] is indeed correct, as other activities (breeding, moult, using certain food types) are non-existent or have a low priority. This task should be solved by the animal in a novel environment, often in a suboptimal habitat, under inclement weather conditions, high level of food competition etc.

The main ecological parameters of a migratory stopover are its duration, fuel deposition rate and energy efficiency. These parameters are interrelated, but the form of relationships is not straightforward. They are studied by the optimal migration theory (Alerstam and Lindström 1990; Weber and Houston 1997;

Houston 1998; Alerstam and Hedenström 1998; Alerstam 2011). To build realistic theoretical models it is necessary to obtain reliable empirical estimates of these main parameters, which appeared to be a non-trivial task.

1.2 Ultimate Factors that Govern Migration

Avian migration has an endogenous basis. It became apparent to the researchers very early that many species, especially long-distance migrants, start their autumn migration rather early, already in summer, long before the feeding conditions deteriorate and cannot any longer support their existence. During the migratory period, obligate migrants enter the so-called migratory disposition (Dolnik 1975; Berthold 1996) which is a special physiological condition, characterised by hyperphagia, deposition of fat, change of the diel pattern of locomotor activity (nocturnal migrants start to be active at night in addition to their normal daytime activity) and spontaneous orientation in the migratory direction. Migrants taken into captivity develop all these key characteristics of the migratory disposition (Kramer 1949; Gwinner 1968), and it should be emphasized that this syndrome is manifested under the constant conditions during many years in captivity (Gwinner 1996). Typically, the period of the rhythm is longer or shorter than 12 months under such constant conditions, attesting to its endogenous circannual nature (Berthold et al. 1972; Gwinner 1986). The amount of nocturnal activity of caged migrants is rather well correlated with the length and sequence of the migratory journey of their free-living conspecifics (Berthold 1973, 1988; Berthold and Leisler 1980), suggesting that nocturnal migratory restlessness of captive migrants (Zugunruhe) is a correct manifestation of the flight activity in the wild. It has also been shown that this amount of nocturnal activity is inherited (Pulido et al. 2001), and when e.g. blackcaps *Sylvia atricapilla* from various populations with different inherited amount of Zugunruhe are crossed in the experiment, their F1 hybrids show an intermediate amount of nocturnal activity (Berthold and Querner 1981). The same is true of the migratory directions: they are also inherited and intermediate in F1 hybrids between different populations (Helbig 1996).

It is generally believed that first-time migrants have an inherited clock-and-compass programme which presupposes that juvenile migrants fly for a certain period of time in a certain direction, then change the direction of their migration and again fly during a certain period of time etc.; finally, they end up in their species-specific winter quarters (Gwinner and Wiltschko 1978; Berthold 1996, 2001). This concept was rather early criticised on theoretical grounds (Rabøl 1985), and in the recent years data has become available suggesting that first-autumn migrants in fact can control their position on the migratory routes from some inherited information on signposts, i.e. some primitive analogue of a map (Fransson et al. 2001; Kullberg et al. 2003, 2007; Freake et al. 2006; Chernetsov et al. 2008a; Liechti et al. 2012). Some satellite tracking data from non-passerine migrants, however, suggest that this inherited map may be actually not so primitive and even comparable with that used by their experienced conspecifics (Gschweng et al. 2008; López-López et al. 2010).

As for experienced migrants, when they return to their breeding areas in spring or migrate to winter quarters after the breeding season, they are believed to have a navigational map and thus to be able to make corrections for displacements, e.g. due to wind drift (Thorup et al. 2007; Chernetsov et al. 2008b).

It can now be regarded a firmly established fact that avian migration has an endogenous basis and that migrating passerines follow their inherited inner urge to migrate that tells them when they should start moving, where they should fly (even though it is not completely clear how it happens and what orientation and navigation mechanisms are in play) and at least in the case of experienced migrants, when they should stop [even though some satellite tracking data from non-passerines strongly suggest that first-time migrants may also have some idea where they are going to and when to stop; Gschweng et al. (2008), López-López et al. (2010)]. However, it is simply not possible to have the sequence of migratory events programmed in every minute detail, because in real life every migrating bird has to deal with too many unpredictable events.

How exactly this general programme of migration is realised, i.e. if the birds accumulate small or large fuel stores, how much time they need for that, if they make short of long migratory flights, how they select stopover habitats, how they behave when at stopover—all these events apparently are not only influenced by the general migratory programme but also by the environmental conditions the migrants encounter, i.e. by the proximate factors. In this book I am trying to give an overview of these organism-environment interactions and look into the factors that play a decisive role during migratory travel.

1.3 Aims of the Study

The goal of is monograph is to put into system the results of my own studies of songbird stopover ecology and behaviour, as well as the existing literature on this subject. I attempted to formulate the main principles that passerine nocturnal migrants obey when selecting a stopover site, when staying there, and when making a decision to resume migratory flight, and the diel timing of flight activity. Knowledge of energy stores at departure and at arrival, together with the new data on flight cost and flight duration may help estimate the distance of single flights and travel rate. It must be kept in mind that these ecological processes happen in the form of behaviour, i.e. habitat selection, spatial and possibly territorial behaviour at stopover. Fuel deposition is only possible if foraging behaviour provides efficient foraging; a migrant should remain vigilant not to be predated etc. Within the framework of the goal of this study the following questions are to be tackled:

- estimating stopover duration and its variation;
- estimating fuel deposition rate and its variation among and between species;

- characterising the relationship between these parameters;
- characterising the mechanism of stopover habitat selection and use by migrants. How important is it to select an optimal habitat if many stopovers are short?
- estimating the role of daytime movements in the progress towards the goal of migration; the role of spatial behaviour at stopovers and identifying the factors that govern it;
- characterising the temporal pattern of flight activity in nocturnally migrating songbirds;
- estimating fuel stores at departure and at arrival and energetic costs of migratory flight
- identifying the main factors that govern the onset and completion of migratory flights.

1.4 The Main Terminology

Under *bird migration* I mean regular seasonal movements of birds between the breeding and the non-breeding parts of their range which include the return of survivors to the breeding area and are endogenously controlled (Salewski and Bruderer 2007; Bruderer et al. 2008; Newton 2008). Migration is an obligate part of the annual cycle except for the non-migratory individuals of partially migrant populations (Terrill and Able 1988; Berthold 1996) and includes the return of all or at least some survivors to the breeding area (Dolnik 1975; Terrill 1990; Bruderer et al. 2008). Even though some authors include nomadic movements (Noskov and Rymkevich 2005) and dispersal (Gauthreaux 1982; Noskov and Rymkevich 2005, 2008; Newton 2008) into the process of migration, or, conversely, treat migration as a form of dispersal (Winker 2000; Nathan et al. 2003; Rappole 2005), neither nomadism nor dispersal are treated as a part of migration in this study.

Migratory stopover is understood here as a stop during the migratory journey that is made for refuelling and/or because continuing migration is temporarily suboptimal for whatever reason (usually due to adverse weather conditions). Migratory stopovers are made for one or several days and usually do not exceed 20–25 days. During stopovers, migrants remain in the *migratory disposition* which is a certain physiological and behavioural condition which characterises a bird ready for migration (Dolnik 1975; Berthold 1996, 2001). Some authors use the word 'stopover' to denote any interruption of migration for any period of time, including for moult or even wintering (Kaiser 1999; Schaub et al. 2005; Yohannes et al. 2005, 2009; Newton 2008; Tøttrup et al. 2012). In this study the words 'migratory stopovers' are used in the narrow sense, and the cases of moult in an area intermediate between breeding and winter range (e.g. Ellegren and Staav 1990) are treated as interruption of migration for moult. The cases of the second breeding cycle in the areas located between the first breeding area and the wintering range recently discovered in North America (Rohwer et al. 2009; Hobson and Robbins 2009) can be called stopovers even less. In such situations a bird is obviously no longer in a migratory disposition but enters the physiological conditions typical of reproduction or moult, which differ significantly from migratory disposition (Gavrilov 1974; Dolnik 1975; Murphy 1991). Borderline cases do occur when it is difficult to decide whether a migrant that interrupted migration for 20–30 days remained in the migratory disposition.

Fuel deposition rate (*FDR*) is the rate of body mass change in a migrant during stopover. It may be measured in grams per hour or grams per day, or as percentage of the initial or lean body mass per hour or day. Even though the word 'deposition' implies mass gain, FDR may be either positive if a bird is gaining body mass, or negative if it loses mass.

Departure fuel load is the amount of fuel (i.e. energy) stored in the form of extra mass (which typically consists of lipids and proteins) deposited above the lean body mass at migratory departure. It has been believed for quite some time that birds and passerines in particular store energy they need for migratory flights (nearly) exclusively as fat (Connel et al. 1960; Odum et al. 1964, 1965; Dolnik 1975; Blem 1976, 1990). However, it has been shown that a portion of energy is stored as wet protein both in waders and geese (McLandress and Raveling 1981; Klaassen et al. 1990; Piersma 1990; Lindström and Piersma 1993) and in songbirds (Klaassen and Biebach 1994; Klaassen et al. 1997; Schwilch et al. 2002).

References

Alerstam T (2003) Bird migration speed. In: Berthold P, Gwinner E, Sonnenschein E (eds) Avian migration. Springer, Berlin

Alerstam T (2011) Optimal bird migration revisited. J Ornithol 152:S5–S25

Alerstam T, Hedenström A (1998) The development of bird migration theory. J Avian Biol 29:343–369

Alerstam T, Lindström Å (1990) Optimal bird migration: the relative importance of time, energy, and safety. In: Gwinner E (ed) Bird migration. Springer, Berlin

Bairlein F (1992) Recent prospects on trans-Saharan migration of songbirds. Ibis 134(Suppl 1):41–46

Barriocanal C, Robson D, Montserrat D (2002) Synoptic situation on long-step night migration by a reed warbler *Acrocephalus scirpaceus* on outward migration. Rev Catalana d'Ornitologia 19:30–34

Begon M, Townsend CR, Harper JL (2006) Ecology: from individuals to ecosystems, 4th edn. Blackwell, Oxford

Berthold P (1973) Relationships between migratory restlessness and migration distance in six *Sylvia* species. Ibis 115:594–599

Berthold P (1988) The control of migration in European warblers. In: Ouellet H (ed) Acta XIX Congr Int Ornithol, vol 1. University of Ottawa Press, Ottawa

Berthold P (1996) Control of bird migration. Chapman and Hall, London

Berthold P (2001) Bird migration: a general survey, 2nd edn. Oxford University Press, Oxford

Berthold P, Leisler B (1980) Migratory restlessness of the Marsh Warbler *Acrocephalus palustris*: a reflection of its unusual migration. Naturwissenschaften 67:472

Berthold P, Querner U (1981) Genetic basis of the migratory behavior in European warblers. Science 212:77–79

Berthold P, Gwinner E, Klein H (1972) Circannuale Periodik bei Grasmücken. I. Periodik des Körpergewichts, der Mauser und der Nachtunruhe bei *Sylvia atricapilla* and *S. borin* unter verschiedenen konstanten Bedingungen. J Ornithol 113:170–190

Biebach H (1990) Strategies of trans-Sahara migrants. In: Gwinner E (ed) Bird migration. Springer, Berlin

Biebach H, Friedrich W, Heine G, Jenni L, Jenni-Eiermann S, Schmidl D (1991) The daily pattern of autumn bird migration in the northern Sahara. Ibis 133:414–422

Biebach H, Biebach I, Friedrich W, Heine G, Partecke J, Schmidl D (2000) Strategies of passerine migration across the Mediterranean sea and the Sahara desert: a radar study. Ibis 142:623–634

Blem CR (1976) Patterns of lipid storage and utilization in birds. Am Zool 16:671–684

Blem CR (1990) Avian energy storage. In: Power DM (ed) Current ornithology, vol 7. Plenum Press, NY

Bolshakov CV (1977) Izuchenie nochnykh migratsiy ptits [metodicheski aspekt] (Study of nocturnal bird migration [methodological aspect]). In: Ilyichev VD (ed) Metody izucheniya migratsiy ptits (Methods of studying bird migration). Nauka, Moscow

Bruderer B (2001) Recent studies modifying current views of nocturnal bird migration in the Mediterranean. Avian Ecol Behav 7:11–25

Bruderer B, Salewski V, Liechti F (2008) Gedanken zur Evolution des Vogelzuges. Orn Beob 105:165–177

Bulyuk V, Chernetsov N (2000) Two migratory flights of sedge warblers *Acrocephalus schoenobaenus* from Finland to Estonia. Ornis Svecica 10:79–83

Chernetsov N (2011) Daytime movements of nocturnal migrants at stopover between two nearby capture sites. J Ornithol 152:1007–1011

Chernetsov N, Mukhin A (2006) Spatial behaviour of European robins *Erithacus rubecula* during migratory stopovers: a telemetry study. Wilson J Ornithol 118:364–373

Chernetsov N, Mukhin A, Ktitorov P (2004) Contrasting spatial behaviour of two long-distance passerine migrants at spring stopovers. Avian Ecol Behav 12:53–61

Chernetsov N, Kishkinev D, Gashkov S, Kosarev V, Bolshakov CV (2008a) Migratory programme of juvenile pied flycatchers, *Ficedula hypoleuca*, from Siberia implies a detour around Central Asia. Anim Behav 75:539–545

Chernetsov N, Kishkinev D, Mouritsen H (2008b) A long-distance avian migrant compensates for longitudinal displacement during spring migration. Curr Biol 18:188–190

Chernetsov N, Kishkinev D, Kosarev V, Bolshakov CV (2011) Not all songbirds calibrate their magnetic compass from twilight cues: a telemetry study. J Exp Biol 214:2540–2543

Cochran WW, Mouritsen H, Wikelski M (2004) Migrating songbirds recalibrate their magnetic compass daily from twilight cues. Science 304:405–408

Connell CE, Odum EP, Kale H (1960) Fat-free weights of birds. Auk 77:1–9

Dolnik VR (1975) Migratsionnoye sostoyanie ptits (Migratory disposition of birds). Nauka, Moscow

Dolnik VR (ed) (1985) Energeticheskie resursy ptits, pereletayuschikh aridnye i gornye prostranstva Sredney Azii i Kazakhstana (Energy resources of birds migrating across arid and mountainous regions of Middle Asia and Kazakhstan). Zoological Institute, Leningrad

Dolnik VR (1990) Bird migration across arid and mountainous regions of Middle Asia and Kasakhstan. In: Gwinner E (ed) Bird migration. Springer, Berlin

Egevang C, Stenhouse IJ, Phillips RA, Petersen A, Fox JW, Silk JRD (2010) Tracking of Arctic terns *Sterna paradisaea* reveals longest animal migration. Proc Natl Acad Sci USA 107:2078–2208

Ellegren H, Staav R (1990) Ruggingsflyttning hos blåhaken *Luscinia s. svecica*. Vår Fågelvärld 49:80–86

Fransson T, Jakobsson S, Johansson P, Kullberg C, Lind J, Vallin A (2001) Magnetic cues trigger extensive refuelling. Nature 414:35–36

Freake M, Muheim R, Phillips JB (2006) Magnetic maps in animals: a theory comes of age? Q Rev Biol 81:327–347

Gauthreaux SA Jr (1971) A radar and direct visual study of passerine spring migration in southern Louisiana. Auk 88:343–365

Gauthreaux SA Jr (1972) Behavioral responses of migrating birds to daylight and darkness: a radar and direct visual study. Wilson Bull 84:136–148

Gauthreaux SA Jr (1982) The ecology and evolution of avian migration systems. In: Farner DS, King JR (eds) Avian biology, vol 6. Academic Press, NY

Gavrilov VM (1974) Metabolizm linyayuschikh ptits (Metabolism of moulting birds). Zool Zhurnal 53:1363–1375

Gill RE Jr, Piersma T, Hufford G, Servranckx R, Riegen A (2005) Crossing the ultimate ecological barrier: evidence for an 11,000 km-long nonstop flight from Alaska to New Zealand and eastern Australia by bar-tailed godwits. Condor 107:1–20

Gill RE Jr, Tibbits TL, Douglas DC, Handel CM, Mulcahy DM, Gottschalck JC, Warnock N, McCaffery BJ, Battley PF, Piersma T (2009) Extreme endurance flights by landbirds crossing the Pacific Ocean: ecological corridor rather than barrier? Proc R Soc Lond B 276:447–457

Gschweng M, Kalko EKV, Querner U, Fiedler W, Berthold P (2008) All across Africa: highly individual migration routes of Eleonora's falcon. Proc R Soc Lond B 275:2887–2896

Gwinner E (1968) Circannuale Periodik als Grundlage des jahreszeitlichen Funktionswandels bei Zugvögeln. Untersuchungen am Fitis (*Phylloscopus trochilus*) und am Waldlaubsänger (*P. sibilatrix*). J Ornithol 109:70–95

Gwinner E (1986) Circannual rhythms in the control of avian migrations. Adv Stud Behav 16:191–228

Gwinner E (1996) Circadian and circannual programmes in avian migration. J Exp Biol 99:39–48

Gwinner E, Wiltschko W (1978) Endogenously controlled changes in migratory direction of the garden warbler, *Sylvia borin*. J Comp Physiol A 125:267–273

Hall-Karlsson KSS, Fransson T (2008) How far do birds fly during one migratory flight stage? Ring Migr 24:95–100

Helbig AJ (1996) Genetic basis, mode of inheritance and evolutionary changes of migratory directions in Palearctic warblers (Aves: Sylviidae). J Exp Biol 199:49–55

Hobson KA, Robbins MB (2009) Origins of late-breeding nomadic sedge wrens in North America: limitations and potential of hydrogen-isotope analysis of soft tissue. Condor 111:188–192

Houston AI (1998) Models of optimal avian migration: state, time and predation. J Avian Biol 29:395–404

Kaiser A (1999) Stopover strategies in birds: a review of methods for estimating stopover length. Bird Study 46(Suppl):S299–S308

Klaassen M, Biebach H (1994) Energetics of fattening and starvation in the long-distance migratory garden warbler, *Sylvia borin*, during the migratory phase. J Comp Physiol B 164:362–371

Klaassen M, Kersten M, Ens BJ (1990) Energetic requirements for maintenance and premigratory body mass gain of waders wintering in Africa. Ardea 78:209–220

Klaassen M, Lindström Å, Zijlstra R (1997) Composition of fuel stores and digestive limitations to fuel deposition rate in the long-distance migratory thrush nightingale, *Luscinia luscinia*. Physiol Zool 70:125–133

Kramer G (1949) Über Richtungstendenzen bei der nächtlichen Zugunruhe gekäfigter Vögel. In: Mayr E, Schüz E (eds) Ornithologie als biologische Wissenschaft. Carl Winter Universitätsverlag, Heidelberg

Krebs JR, Davies NB (eds) (1991) Behavioural ecology: an evolutionary approach, 3rd edn. Blackwell, Oxford

Kullberg C, Lind J, Fransson T, Jakobsson S, Vallin A (2003) Magnetic cues and time of season affect fuel deposition in migratory thrush nightingales (*Luscinia luscinia*). Proc R Soc Lond B 270:373–378

Kullberg C, Henshaw I, Jakobsson S, Johansson P, Fransson T (2007) Fuelling decisions in migratory birds: geomagnetic cues override the seasonal effect. Proc R Soc Lond B 274:2145–2151

Liechti F, Komenda-Zehnder S, Bruderer B (2012) Orientation of passerine trans-Sahara migrants: the directional shift ('Zugknick') reconsidered for free-flying birds. Anim Behav 83:63–68

Lindström Å, Piersma T (1993) Mass changes in migrating birds: the evidence for fat and protein storage re-examined. Ibis 135:70–78

López-López P, Limiñana R, Mellone U, Urios V (2010) From the Mediterranean Sea to Madagascar: Are there ecological barriers for the long-distance migrant Eleonora's falcon? Landscape Ecol 25:803–813

Lyuleeva DS (1970) Energiya poleta u lastochek i strizhey (Energy of flight in swallows and swifts). Doklady Acad Sci USSR 190:1467–1469

McLandress MR, Raveling DG (1981) Changes in diet and body composition of Canada geese before spring migration. Auk 98:65–79

Moore FR (1990) Evidence for redetermination of migratory direction following wind displacement. Auk 107:425–428

Muheim R, Phillips JB, Åkesson S (2006) Polarized light cues underlie compass calibration in migratory songbirds. Science 313:837–839

Muheim R, Åkesson S, Phillips JB (2007) Magnetic compass of migratory Savannah sparrows is calibrated by skylight polarization at sunrise and sunset. J Ornithol 148(Suppl 2):S485–S494

Muheim R, Phillips JB, Deutschlander ME (2009) White-throated sparrows calibrate their magnetic compass by polarized light cues during both autumn and spring migration. J Exp Biol 212:3466–3472

Murphy ME (1991) Energetics and nutrition of molt. In: Carey C (ed) Avian energetics and nutritional ecology. Chapman and Hall, NY

Nathan R, Perry G, Cronin JT, Strand AE, Cain ML (2003) Methods for estimating long-distance dispersal. Oikos 103:261–273

Newton I (2008) The migration ecology of birds. Academic Press, London

Noskov GA, Rymkevich TA (2005) Formy migratsionnoy aktivnosti v godovom tsikle ptits (Forms of migratory activity in the annual cycle of birds). In: Iovchenko NP (ed) Ornitologicheskie issledovaniya v Priladozhye (Ornithological research in the Ladoga area). St. Petersburg University Press, St. Petersburg

Noskov GA, Rymkevich TA (2008) Migratsionnaya aktivnost' v godovom tsikle vorobyinykh ptits i formy ee proyavleniya (Migratory activity in the annual cycle of passerines and the forms of its manifestation). Zool Zhurnal 87:446–457

Odum EP, Rogers DT, Hicks DL (1964) Homeostasis of the nonfat components of migrating birds. Science 143:1037–1039

Odum EP, Marshall SG, Marples TG (1965) The caloric content of migrating birds. Ecology 46:901–904

Paxton KL, Van Riper C III, O'Brien C (2008) Movements patterns and stopover ecology of Wilson's warblers during spring migration on the lower Colorado river in southwestern Arizona. Condor 110:672–681

Piersma T (1990) Pre-migratory 'fattening' usually involved more than the deposition of fat alone. Ring Migr 11:113–115

Piersma T, Pérez-Tris J, Mouritsen H, Bauchinger U, Bairlein F (2005) Is there a 'migratory syndrome' common to all migrant birds? Ann NY Acad Sci 1046:282–293

Pulido F, Berthold P, Mohr G, Querner U (2001) Heritability of the timing of autumn migration in a natural bird population. Proc R Soc Lond B 268:953–959

Rabøl J (1985) The moving goal area and the orientation system of migrant birds. Dansk Orn Foren Tidsskr 79:29–42

Rabøl J (2010) Orientation by passerine birds under conflicting magnetic and stellar conditions: no calibration in relation to the magnetic field. Dansk Orn Foren Tidskr 104:85–102

Rappole JH (2005) Evolution of Old and New World migration systems: a response to Bell. Ardea 93:125–131

Rohwer S, Hobson KA, Rohwer VG (2009) Migratory double breeding in Neotropical migrant birds. Proc Natl Acad Sci USA 106:19050–19055

Salewski V, Bruderer B (2007) The evolution of bird migration—a synthesis. Naturwissenschaften 94:268–279

Schaub M, Kania W, Köppen U (2005) Variation of primary production during winter induces synchrony in survival rates in migratory white storks *Ciconia ciconia*. J Anim Ecol 74:656–666

Schmaljohann H, Liechti F, Bruderer B (2007) Songbird migration across the Sahara: the non-stop hypothesis rejected! Proc R Soc Lond B 274:735–739

Schwilch R, Grattarola A, Spina F, Jenni L (2002) Protein loss during long-distance migratory flight in passerine birds: adaptation and constraints. J Exp Biol 205:687–695

Terrill SB (1990) Evolutionary aspects of orientation and migration in birds. Experientia 46:395–404

Terrill SB, Able KP (1988) Bird migration terminology. Auk 105:205–206

Thorup K, Bisson I-A, Bowlin MS, Holland RA, Wingfield JC, Ramenofsky M, Wikelski M (2007) Evidence for a navigational map stretching across the continental U.S. in a migratory songbird. Proc Natl Acad Sci USA 104:18115–18119

Titov N (1999) Home ranges in two passerine nocturnal migrants at a stopover site in autumn. Avian Ecol Behav 3:69–78

Tøttrup AP, Klaassen RHG, Strandberg R, Thorup K, Kristensen MW, Jørgensen PS, Fox J, Afanasyev V, Rahbek C, Alerstam T (2012) The annual cycle of a trans-equatorial Eurasian-African passerine migrant: different spatio-temporal strategies for autumn and spring migration. Proc R Soc B 279:1008–1016

Wang Y, Moore FR (1993) Relation between migratory activity and energetic condition among thrushes (*Turdinae*) following passage across the Gulf of Mexico. Condor 95:934–943

Ward MP, Raim A (2011) The fly-and-social foraging hypothesis for diurnal migration: why American crows migrate during the day. Behav Ecol Sociobiol 65:1411–1418

Weber TP, Houston AI (1997) A general model for time-minimising avian migration. J Theor Biol 185:447–458

Wikelski M, Tarlow EM, Raim A, Diehl RH, Larkin RP, Visser GH (2003) Costs of migration in free-flying songbirds. Nature 423:704

Williams TC, Williams JM (1990) The orientation of transoceanic migrants. In: Gwinner E (ed) Bird migration. Springer, Berlin

Winker K (2000) Migration and speciation. Nature 404:36

Yohannes E, Hobson KA, Pearson DJ, Wassenaar LI (2005) Stable isotope analysis of feathers help identify autumn stopover sites of three long-distance migrants in northeastern Africa. J Avian Biol 36:235–241

Yohannes E, Biebach H, Nikolaus G, Pearson DJ (2009) Passerine migration strategies and body mass variation along geographic sectors across east Africa, the Middle East and the Arabian Peninsula. J Ornithol 150:369–381

Chapter 2
Stopover Duration

Abstract In this chapter I describe and critically discuss the methods of estimating stopover duration. Potential pitfalls of the capture-mark-recapture analysis are discussed, and radio telemetry is recommended. However, an important benefit of capture-mark-recapture models is that they not only provide estimates of the mean stopover duration, but often allow studying within-species diversity of stopover strategies. The empirical results from the literature are reviewed. It is concluded that the duration of migratory stopovers of songbirds usually varies between 1 and 15 days. Sometimes, especially before and just after crossing large ecological barriers (large water bodies, deserts) stopovers may be longer and reach 20–25 days, occasionally even longer. Significant proportions of migrants stop over for one day only and continue migration on the first night after arrival.

2.1 Methods of Estimating Stopover Duration

2.1.1 Minimum Stopover Length Estimates

Arguably, the most straightforward way to estimate stopover duration is to set up a trapping site, to capture stopping migrants, mark and try to recapture or resight them to read the marks. As most songbirds are inconspicuous, recaptures are used more frequently than resightings, even though it has been recently demonstrated that using both methods of encountering marked individuals significantly enhances the accuracy of stopover duration estimates as compared with using recaptures only (Salewski et al. 2007; Verkuil et al. 2010). The reason for this may be the fact that recapture probability does not remain constant throughout the stopover period, but generally declines towards its end (see below).

Large-scale capture-mark-recapture studies have become possible since the 1960s when mist-nets came into common usage for trapping birds. The simplest method to estimate the duration of migratory stopovers on the basis of

N. Chernetsov, *Passerine Migration*, DOI: 10.1007/978-3-642-29020-6_2,

capture-recapture data is the so-called *minimum stopover length, MSL*. Minimum stopover length (or duration) is time in days (or in hours) elapsed between the first capture of a bird and its last capture or resighting. As nocturnal migrants normally do not perform migratory flights during the daytime, their MSL is usually measured in whole numbers of days. Exceptions to this rule are the stopover durations of passerines that arrive after crossing a large water body (e.g. Gulf of Mexico; Gauthreaux 1971; Wang and Moore 1997), when flights may extend into the daytime and thus the first stopover day may not be a full one. It should also be noted that a nocturnal migrant that was only captured (once or multiple times) within a single day has a stopover duration of one, not zero days; a bird captured in two consecutive days has MSL $= 2$ days, etc.

When using MSL as a stopover duration estimate, it is assumed that time spent at stopover before the first capture and after the last capture is negligible. Even though under some circumstances this assumption may be justified (especially its former part), generally speaking, it is a very strong assumption. To know the true stopover duration, it is necessary to estimate its hidden part, i.e. time spent at stopover before the first capture and after the last capture or resighting. Therefore, MSL underestimates stopover duration, as apparent already from the word 'minimum'. However, most authors only include birds captured at least twice in the analysis, and neglect individuals that were never recaptured (which are a vast majority in most cases). Obviously, this may lead to overestimates of the mean stopover duration. Therefore, MSL generally provides biased estimates, and it is even unclear whether these estimated are biased high or low.

2.1.2 Capture-Mark-Recapture Models: Estimating the Expected Stopover Duration

To solve this problem, it has been suggested that stochastic capture-mark-recapture models should be used that are common in animal demography studies (Lavée et al. 1991; Kaiser 1993, 1995, 1999). These models are used to estimate survival rates (Lebreton et al. 1992; Lebreton 2001; Burnham and Anderson 2002; Williams et al. 2002). They are used to compensate for imperfect capture probability. If all individuals present at the study site were captured every day with a 100% probability, we would have known their arrival and departure dates exactly and there would have been no reason to construct any models.

The main idea of this method is that all the birds present at the study site are considered to be an open population where all changes of numbers are due to immigration and emigration. Indeed, daily mortality rate at stopovers is usually low. On the basis of many individual capture histories (recorded as e.g. 10011011, where a 1 is a day when this individual was captured and 0 is a day when it was not captured) the daily apparent survival rate and daily capture probability are estimated. Capture probability (p) is the probability that a bird present at the study site is captured on a given day. Apparent, or local, survival rate (φ) is the probability that a bird present in the population on day i will be present on day $i + 1$. The local survival rate is the product of the true survival rate and the probability of staying in the study area. As the daily true survival rate is usually very close to 1, within the context of migratory stopovers φ is often called just the probability of stay. Emigration probability is $1 - \varphi$. Obviously, as both p and φ are probabilities, they vary between 0 and 1.

To identify the model that fits the data best (i.e. describes the dataset in the best possible manner), a number of models based on different assumptions are fitted. We may assume that φ, or p, or both parameters are time-dependent, i.e. within the context of migratory stopovers vary from day to day of trapping. Or we may assume that these parameters are time-independent, i.e. constant. Or they may vary between e.g. sex and age groups, or depend on the time elapsed since marking etc. All these models are theoretical ones and are based on our *a priori* knowledge of the ecology and behaviour of the animals studied. When many models are constructed, they are compared to find the most parsimonious one(s), i.e. the model(s) that explain the data best when using the smallest number of parameters. Models are usually compared by Akaike's Information Criterion (AIC) which is a measure of deviation of the model from the data, taking into consideration the number of parameters (Burnham and Anderson 2002). The model which fits the data best is the one that has the lowest AIC value (obviously, it depends on the models that we included into our comparison). The difference of a certain model from the best one is denoted as ΔAIC. In the best model $\Delta\text{AIC} = 0$ by definition.

It may happen that AIC values for two or several models are very similar, so that ΔAIC of some models is very low. This means that these models fit the data (nearly) equally well and it is difficult to say which one is better. In such case AIC weights are calculated that are probabilities of that the given model fits the data best. It is usually assumed that models with AIC weights >0.05 are worth considering. Usually AIC values are close in models with very similar structure. It suggests that the model structure is a correct one.

The expected stopover duration after the first capture is estimated as $S_a = -1/\ln \varphi$ (Seber 1982). This formula is a particular case of the following more general expression (Schaub et al. 2001):

$$S_a = (1 - \varphi_1) \cdot (-1/\ln \varphi_1) + \varphi_1 \cdot (1 - \varphi_2) \cdot (-1/\ln \varphi_2) + \varphi_1 \cdot \varphi_2 \cdot (1 - \varphi_3) \cdot (-1/\ln \varphi_3) + \cdots \tag{2.1}$$

when $\varphi = \text{const}$ (φ_1 is the probability of stay between the first and the second day; φ_2 between the second and the third day etc.).

Formula 2.1 makes it possible to obtain estimates of the expected stopover duration after the first capture. However, a bird may have arrived to the study site some time before the first capture and remained undetected. Therefore it is necessary to estimate this hidden stopover time before the initial capture. This is done by analysing the inverted capture histories when day 1 becomes day n, day 2, day n − 1 etc.; day n becomes day 1. From the inverted capture histories, seniority (γ) is estimated, which is the probability that a bird present in the population on day

i was also present on day $i - 1$ (Pradel 1996; Pradel et al. 1997). Immigration probability is $1 - \gamma$. Estimates of seniority allow calculation of stopover duration estimates before the first capture (S_b) similar to S_a calculation. The total stopover duration before and after the first capture is estimated as $S = S_a + S_b = (-1/\ln \varphi) + (-1/\ln \gamma)$ (Schaub et al. 2001).

It has been suggested that the term involving γ, i.e. S_b estimate, is superfluous and should not be included into the estimate of the total stopover duration (Efford 2005). It is so because presence at the time moment i is conditional on having not departed from arrival to i, but the birds that had arrived at e.g. $i - 2$ and departed before i are not included in calculations, biasing the estimate high. This is a matter of some debate (Efford 2005; Pradel et al. 2005), but most authors currently prefer not to include S_b into the total stopover duration estimate (Salewski et al. 2007; Bayly and Rumsey 2007; Chernetsov et al. 2007, Chernetsov 2010).

In practice this methodological problem is less severe than it looks. It has been mentioned before that in some cases the assumption that stopover migrants are captured quite soon after their arrival is justified. For instance, on the Courish Spit (south-eastern Baltic coast), like in many other coastal areas, passage has a pronounced wave-like pattern (Floerike 1893; Blyumental et al. 1967; Dolnik 1975; Titov and Chernetsov 1999). This wave-like pattern of passage influences stopover behaviour: periods with few individuals at stopover, many of which are captured repeatedly, alternate with peaks (or waves) of captures when many previously unmarked birds are trapped. It seems obvious that on the day of a wave of migration most freshly marked birds are new arrivals.

This intuitive idea is supported by our seniority analysis: on the first day of a migratory wave the probability that newly captured European robins *Erithacus rubecula* have been present at stopover earlier undetected is very low, as indicated by low seniority estimates. On a quiet day, i.e. during a pause between two peaks of captures, seniority is high (Table 2.1). It should be also kept in mind that the movement rate of stopover migrants, which governs capture probability to a large extent, is generally the highest during the first and sometimes the second day upon arrival and strongly declines afterwards (see Chap. 6).

A serious problem of using capture-mark-recapture models for estimating stopover duration is that this method is very data-hungry (Chernetsov and Titov 2000). Because of that, time-dependent models often have higher AIC values (e.g. lower support) than less sophisticated and often less realistic models with constant (time- and time-since-marking-independent) parameters. It happens because of overfitting, i.e. the number of parameters to estimate is too large for the number of capture histories available (Burnham and Anderson 2002). To solve this problem, data may be pooled, i.e. several days of trapping are treated as one capture event (Schaub and Jenni 2001a; Schaub et al. 2001). However, if more than one capture happens within this pooled capture event, information is lost. Moreover, data pooling has been shown to bias parameter estimates (Hargrove and Borland 1994; Morris et al. 2005b). To circumvent this problem, it has been suggested that the models should be fitted with the parameters fixed for several consecutive days (multiple day constancy models; Morris et al. 2005a). The number of parameters to

Table 2.1 Parameters of capture-mark-recapture models that describe stopover duration of European robins in Rybachy (Courish Spit, Eastern Baltic) in autumn 1996 (from Titov and Chernetsov 1999, revised)

Parameters		Mean estimate	95% confidence interval
Direct capture histories	φ_{11}	0.478	0.399–0.559
	φ_{12}	0.243	0.202–0.288
	φ_{13}	0.131	0.110–0.156
	φ_2	0.521	0.450–0.591
	φ_3	0.802	0.767–0.833
	p_1	0.266	0.237–0.297
	p_2	0.195	0.159–0.237
Inverted capture histories	γ_{11}	0.892	0.665–0.972
	γ_{12}	0.358	0.295–0.427
	γ_{13}	0.116	0.098–0.138
	γ_2	0.553	0.476–0.629
	γ_3	0.772	0.743–0.799
	p	0.236	0.212–0.262

estimate is reduced, as during data pooling, but the information on multiple captures within this time interval is not wasted. Therefore, using multiple day constancy models can be recommended when the data is scarce.

Another simple method to estimate the true stopover duration has been suggested by Vysotsky (1998). This method is based on the assumption that the first capture happens on average in the middle of stopover, and the last capture in the middle of the period between the first capture and departure. Therefore, MSL is on average 25% of the true stopover duration. Vysotsky's estimate is based on the implicit assumption that capture probability is constant throughout the period of stopover. This assumption ignores the known patterns of spatial behaviour at stopover (see Chap. 6) and is not justified. As all the capture methods used in standardised trapping projects are based on the passive capture of moving migrants, the capture probability is heavily dependent on the bird's mobility. The scale of movements strongly changes during stopover: it usually declines, often quite significantly (Chernetsov 2005; Chernetsov and Mukhin 2006).

2.1.3 Estimating the Proportions of 'Transients' and 'Non-Transients'

It should be emphasized that capture-mark-recapture models not only allow estimates of the mean expected stopover duration, but also to study the diversity of stopover strategies. As shown by our data (Chernetsov 1998b; Panov and Chernetsov 2010; Table 2.2; Sect. 2.4) and that of other authors (Salewski and Schaub 2007), the empirical data are often best described by a time-since-marking model (Cooch and White 2005), often called age-dependent models for brevity.

This usage originates from demographic studies where stochastic capture-mark-recapture models were initially developed and used and where capture events are often breeding seasons, so that the time since marking is often measured in years and thus shows the age of animals. In the context of migratory stopover 'age' is measured in days elapsed since the first capture.

Support of such models usually means that migrants that have just arrived with a high probability leave the stopover area after a one-day stopover (during the first night following arrival). The birds that stay for the second and even for the third day are likely to remain longer, usually for 5–8 days or more. In such situations it is justified to estimate the proportion of transients that make one-day stopovers and the mean expected stopover duration of non-transients, and not to pool both groups. The probability that a newly trapped individual is a transient (i.e. has a zero probability to remain longer than one day; Pradel et al. 1997) and stopover duration of non-transients is estimated from time-since-marking models. The probability that a fresh capture is a transient is calculated as $\tau = 1 - \varphi_1/\varphi_2$, where φ_1 is the probability of stay estimate in the first 'age' class (i.e. between the first and the second days of stopover), and φ_2 is the probability of stay estimate in the second class (between the subsequent days; Salewski et al. 2007).

Making distinction between transients and non-transients and estimating their stopover duration is not only justified by the formal logic of capture-mark-recapture modelling (Pradel et al. 1997; Salewski et al. 2007), but also by physiology-based motivation to continue migration (Rappole and Warner 1976; Tsvey et al. 2007). Certainly, strictly speaking, all migrants at stopover are transients, and it is not correct to call birds that just stay a little longer non-transients. However, making distinction between 'transients' (or 'flyers', birds that stop for one and sometimes two days) and non-transients (or 'feeders'; Rappole and Warner 1976) that stay for a longer period is justified. This usage may be followed in order to keep capture-mark-recapture terminology uniform and mutually intelligible between different applications of this analysis.

2.1.4 Method of Elevated Mist-Nets

It should be emphasized that capture-mark-recapture models do not allow measurements of stopover duration of individual migrants. They only produce estimates of the mean expected stopover duration of cohorts of birds and confidence intervals of the means. However, in many cases it is essential to know the individual durations of stay. It is very difficult, nearly impossible, to observe the moment of arrival (but see Chap. 7). However, as mentioned above, on the first day of the peak of passage most trapped birds are indeed fresh arrivals (Table 2.1). High seniority values of migrants initially captured during the pauses of passage together with low seniority on the first day of migratory waves strongly suggest that birds first captured between the peaks of captures must have arrived with the previous wave of arrivals.

Fig. 2.1 Elevated mist-nets at the trapping site Rybachy on the Courish Spit. Photo by E. Popov

At the Biological Station Rybachy we have developed a method of capturing passerine nocturnal migrants during take-off in mist-nets set above the canopy (Fig. 2.1). It has been shown that captures in such nets that occur between 60 min after sunset and 60 min before sunrise do not refer to local movements that diurnal passerines do not perform in the darkness, as shown e.g. by the complete lack of their captures in standard 'low' mist-nets. Nocturnal captures in elevated, or 'high', mist-nets only refer to nocturnal migratory flights: take-offs, landing, or very low altitude migration (Bolshakov et al. 2000; Bulyuk 2006; Bulyuk and Tsvey 2006). When birds captured in nigh mist-nets have been previously marked (ringed) during their daytime stopover movements at the study site it is possible to claim that we exactly know the time of their migratory departure from stopover (Bolshakov et al. 2003a, b; Bulyuk and Tsvey 2006).

It should be mentioned that nocturnal trapping in high mist-nets is a very labour-consuming activity, especially because it is only useful when run in parallel with a large-scale daytime trapping project in the same area. In Rybachy, the project was run with more than 700 m of mist-nets checked hourly 24 h a day during seven autumn migratory seasons. It resulted in 78 documented migratory departures of European robins (the most common species; Bulyuk and Tsvey 2006) and lower number of captures of other songbird species. The minimum stopover duration of European robins varied between 1 and 12 days, on average 4.0 days (SD = 2.84, n = 78). However, if we disregard these additional, non-standard captures at take-off, and just analyse captures of the same individuals in the standard nets, their mean stopover duration would be estimated at 1.8 days (SD = 1.66), with variation between 1 and 9 days. The difference between these two estimates in individual birds is 0–10 days, or 0–91%, on average 46%. Apparently, this difference is too large to be ignored, and estimates based on recaptures in standard mist-nets are seriously low biased.

2.1.5 Radio-Tagging

The most objective method of estimating stopover duration is radio-tagging of migrants by small VHF transmitters. This method allows tracking of individual migrants until their migratory departure. Radio-tracking of passerines at migratory stopovers used to be hampered by the small size of the birds. Only in the 1990s, when small tags with a mass below 1 g became commercially available, radio-tagging studies of songbird migrants started to be published (e.g. Aborn and Moore 1997).

Lightweight transmitters used for tagging small (10–20 g) passerines usually have a limited signal strength and, as a result, not very large reception range, usually ca. 400–1,500 m depending on the habitat structure. Because of that, tagged birds that move even not very long distances may be lost. It should be kept in mind that even birds that normally remain within a rather limited home range during stopover, like European robins (Titov 1999a, b; Lajda 2001), may move for up to 1.5–2 km during the daytime (Chernetsov and Mukhin 2006). In order not to mistake such relatively long-distance diurnal movements for migratory departures, it is necessary either to do research on small islands where movement opportunities are limited (Aborn and Moore 1997; Schmaljohann et al. 2011), or to keep the birds under surveillance 24 h and to detect the moment of migratory take-off exactly (Chernetsov et al. 2004; Chernetsov and Mukhin 2006; Mills et al. 2011). This is especially important when species like e.g. the pied flycatcher *Ficedula hypoleuca* or garden warblers *Sylvia borin* are studied that are capable of moving for up to several kilometres during migratory stopovers (Chernetsov et al. 2004; Fransson et al. 2008). The point is that not every case when the signal from a tagged bird is no longer received from the ground can be automatically believed to indicate migratory departure. Exploratory flights or landscape-scale movements may result in the loss of signal (Schmaljohann et al. 2011; Mills et al. 2011).

One of the main strengths of radio telemetry is that detection probability of a bird is not dependent on its mobility. This makes telemetry studies radically different from capture-mark-recapture data when detection (recapture or resighting) probability is strongly dependent on spatial behaviour of migrants that may significantly change in the course of stopover. Therefore if the telemetry data is collected carefully, it should be regarded as the most accurate estimate of stopover duration. On the other hand, due to the high cost of transmitters as compared with rings, and due to labour intensity of ground-based telemetry (however, the advent of automatic receiving units helps to solve this problem) the sample size of radio-tagging studies will always remain more limited than in the ringing-based studies. Finally, as already mentioned, capture-mark-recapture statistics yields mean estimates for cohorts of birds and confidence limits for these estimates. Radio-tagging, like captures in elevated mist-nets, allows measurements of stopover duration of individual migrants.

2.2 Estimates Based on Re-Encounters of Marked Birds

2.2.1 *Biased Estimates: Migratory Stopovers, Postfledging Movements, and Moult*

The mean values of stopover duration estimates based on capture-mark-recapture data vary broadly. For instance, the stopover duration of Eurasian reed warblers *Acrocephalus scirpaceus* in Bolle di Magadino (Switzerland) was estimated as 12.25 days (SD = 1.67; n = 567) which is significantly longer than the MSL = 6.0 days (range 2–34 days, SD = 5.04; n = 108; Schaub et al. 2001).

These estimates deserve special attention. Duration of stay for 34 days in Central Europe can hardly refer to migratory stopovers. Beyond doubt they refer to the duration of stay of birds in juvenile moult that have not yet started autumn migration. Estimates of stopover duration during autumn passage are biased towards longer stopovers, especially in the Eurasian reed warbler, blackcap and other species with pronounced juvenile moult of body feathers (Jenni and Winkler 1994), because Schaub and Jenni (2000, 2001b) believe that autumn migration broadly overlaps with juvenile body moult. As supporting evidence these authors cited Herremans (1990) who tape-lured moulting Eurasian reed warblers at night in Belgium. However, Marc Herremans did not claim that the birds he captured were overlapping moult and autumn migration; he did not rule out that they were performing pre-migratory movements (juvenile dispersal).

Studies by Bulyuk et al. (2000) and Mukhin (2004) and Mukhin et al. (2005) clearly showed that juvenile Eurasian reed warblers performed nocturnal postfledging movements at the age of 36–50 days, i.e. before the onset of autumn migratory movements towards their winter quarters. The range of such movements is probably ca. several dozens of kilometres which is typical of juvenile dispersal distances of *Acrocephalus* warblers (Grüll and Zwicker 1981; Chernetsov 1998a; Paradis et al. 1998). These movements are not generally directed towards the winter quarters and are not migratory but postfledging movements of birds that are not in migratory disposition (Mukhin 2004; Mukhin et al. 2005). During these movements, Eurasian reed warblers and some other songbirds can be attracted by playback of conspecific and heterospecific song (Bulyuk et al. 2000; Mukhin et al. 2008). Therefore, the opinion that hatching-year Eurasian reed warblers broadly overlap juvenile moult and autumn migration in Central Europe is erroneous. Conversely, their long stays in the study sites (significantly longer in moulting individuals than in their conspecifics that have completed moult) found by Schaub and Jenni (2000, 2001b) strongly suggest that these are local birds in the broad meaning that are performing local postfledging movements and have not yet started autumn migration. Apparently, the inclusion of such birds in analysis biases stopover length estimates high (and FDR estimates low, see Chap. 3).

2.2.2 Reliable Capture-Mark-Recapture Estimates of Stopover Duration

Michael Schaub and Lukas Jenni estimated stopover duration of Eurasian reed warblers, sedge warblers *Acrocephalus schoenobaenus* and garden warblers at many sites from Finland to Spain and northern Africa by capture-mark-recapture models (Schaub and Jenni 2001a). The data were taken from the database of the European-African Songbird Migration Network (1994–1996; Bairlein 1995, 1998). For Eurasian reed warblers, I only used estimates for non-moulting individuals, since only they refer to the migrating birds (see above).

The mean stopover duration of Eurasian reed warblers was 3.7–7.7 days before the first capture and 3.1–6.1 days after the first capture (Table A1 in Schaub and Jenni 2001a; see Sect. 2.1.2 for the disputed usefulness of the estimates of stopover length before the first capture). Long stopovers (>10 days if the period before the first capture is included) were recoded on the Iberian Peninsula and in Morocco, and also in Bolle di Magadino (Switzerland) and on Lake Galenbeck in NE Germany.

Realistic estimates for the sedge warbler (as opposed to the obviously erroneous estimate for the Norwegian site, see above) are 1.8–12.0 days, obtained in Illmitz (E Austria), Rauvola (Finland), Rybachy (Kaliningrad Region), Lake Galenbeck and in Bolle di Magadino (0.9–6.3 days when only estimates after the first capture were included; Table A2 in Schaub and Jenni 2001a). In Illmitz the estimates varied between the years between 1.8 and 8.8 days. The authors suggest that such a broad annual variation was due to varying abundance of plum aphids *Hyalopterus pruni* that is believed to be the main food of sedge warblers in late summer and autumn (Bibby and Green 1981).

Garden warblers stopped for 3.4–14.1 days (1.7–7.1 days when only estimates after the initial capture are included; Table A3 in Schaub and Jenni 2001a). The very long estimate of 20.7 days refers to the total stopover duration in Ginak in Gambia which may refer to the wintering range rather than autumn passage.

It should however be emphasized that not all long stopovers should automatically be dismissed as unrelated to 'true' migration. This is shown by many estimates of stopover duration of passerine migrants that are facing ecological barriers and therefore need to accumulate significant fuel stores or have just crossed a barrier and need to recover.

Long-term data from a site just north of a major barrier are available from Eilat, an Israeli site on the northern tip of Gulf of Aqaba of the Red Sea. I analysed the

17-year data from 1984–2001 (except of 1987), and found that just 55 individuals stopped by at least two days (MSL), of 356 first captures. Of these 55 birds, only four were recaptured more than 10 days after the initial capture. However, in spring, after crossing the Sahara, 357 sedge warblers were recaptured at least one day after the first capture, 54 of them after at least 10 days. The maximum duration of stay in Eilat in spring was 46 days. Even if this stay referred to a sick or otherwise abnormal bird, durations of stay of 33, 35, 36 days probably represent the actual stopovers of sedge warblers during spring passage in Eilat. The body mass of these birds increased on average by 14.4%, whereas body mass increase in sedge warblers that stopped for 2–10 days comprised 7.3% (t-test, $t = 3.23$, $df = 314$, $p < 0.002$). Therefore, stopovers longer than 10 days are made by normal migrating sedge warblers, and not by sick or abnormal individuals. Eurasian reed warblers also make long stopovers in Eilat in spring, with MSL of up to 28 days. Of 270 individuals that stopped for more than one day, 27 remained for more than 10 days and increased their body mass on average by 12%.

Blackcaps stopped over in Eilat in spring for 3.54 days (males, $SD = 3.03$, $n = 386$) and for 3.86 days (females, $SD = 3.17$, $n = 526$), with significant sex-related difference in MSL (Mann–Whitney test, $U = 92460$, $p = 0.025$; Yosef and Wineman 2010). The maximum MSL was up to 35 days, i.e. similar to the values found in *Acrocephalus* warblers.

These estimates are made by the MSL method which has multiple methodological issues (Sect. 2.1.1). However, as we are here most interested not in obtaining unbiased estimates of the mean values (which is problematic by the MSL method), but in checking the validity of maximum estimates, the use of this method is justified in this case. It is also worth noting that these long stopovers were recorded during spring passage, when migrants are believed to be travelling fast because there is a high premium in arriving early to the breeding grounds (Kokko 1999, but see Forstmeier 2002).

Stopovers of Eurasian reed warblers before and after crossing the Sahara may be quite long, too. In autumn in Sidi Bou Ghaba near the Moroccan Atlantic coast adults stopped on average for 12.4 days, hatching-year birds for 17.8 days (capture-mark-recapture estimates with time before initial capture included; Rguibi-Idrissi et al. 2003). In Kerbacha on the Mediterranean coast of Morocco the values were 7.5 and 12.9 days, respectively. In spring the mean stopover duration estimates in Sidi Bou Ghaba were 8.1 days for adults and 13.5 days for yearlings; in Kerbacha the values were 3.2 days and 8.6 days, respectively. It should be however noted that ageing of Eurasian reed warblers during spring migration is not unproblematic.

Our estimates of the mean stopover duration in six species of passerine nocturnal migrants in Dzhanybek (northern part of the Caspian plain) varied between 0.85 days in the willow warbler to 3.4 days in the garden warbler in 2003 (Table 2.2). Even the upper limit of the 95% confidence interval did not exceed 5.7 (in the red-breasted flycatcher *Ficedula parva* in 2003)—5.8 days (in the garden warbler in 2003). It should be however, noted that unlike Schaub and Jenni (2001a) and Rguibi-Idrissi et al. (2003), we did not include stopover duration before initial capture in our estimates, following Efford (2005).

Table 2.2 Parameters of CMR models and stopover duration in songbird migrants in Dzhanybek (Western Kazakhstan)

Species, year	Proportion of transients	φ_1, mean and 95% confidence interval	φ_2 (or φ), mean and 95% confidence interval	Stopover duration, days (mean and 95% confidence interval)
Garden warbler 2003	0.58	0.313; 0.182–0.482	0.746; 0.620–0.841	3.41 (2.09–5.77)
Garden warbler 2004			0.572; 0.443–0.693	1.79 (1.23–2.73)
Common redstart 2003			0.576; 0.383–0.748	1.81 (1.04–3.44)
Common redstart 2004	0.80	0.139; 0.080–0.231	0.684; 0.575–0.776	2.63 (1.81–3.94)
Chiffchaff 2004			0.695; 0.564–0.800	2.75 (1.75–4.48)
Willow warbler 2004			0.307; 0.083–0.685	0.85 (0.40–2.64)
European robin 2004	0.65	0.215; 0.128–0.338	0.614; 0.424–0.774	2.05 (1.17–3.90)
Red-breasted flycatcher 2003			0.712; 0.540–0.839	2.94 (1.62–5.70)
Red-breasted flycatcher 2004	0.55	0.221; 0.109–0.396	0.490; 0.378–0.602	1.40 (1.03–1.97)

At the same site, in the spring of 2005 just two species of songbird migrants, the Blyth's reed warbler *Acrocephalus dumetorum* and garden warbler, were captured in significant numbers. The scarcity of recaptures did not allow estimates of stopover duration at spring passage, but the very fact that of 110 garden warblers marked in spring not a single bird was recaptured on subsequent days (in autumn this proportion was 43 out of 599, 7%) strongly suggests that the bulk of birds made one-day stopovers. Thus, after crossing the arid areas northwest of the Caspian Sea (garden warblers) and the deserts of Central Asia in spring (Blyth's reed warblers) migrants did not need to make prolonged stopovers to recover. A similar situation was recorded after spring crossings of the Gulf of Mexico (e.g. Wang and Moore 1993; Aborn and Moore 1997). It may be mentioned that MSL of garden warblers before crossing the Sahara in Amurum (Nigeria) in spring varied between 1 and 13 days, on average 5.0 days (SE = 1.15; n = 11, calculated from Fig. 5 in Ottosson et al. 2005).

In southern France the mean duration of migratory stopovers of Eurasian reed warblers varied between 6.1 days in late July and 11.1 days in late October, on average 8.5 days (Balança and Schaub 2005). This is very close to the estimates for this species at other European sites (Schaub and Jenni 2001a). In southern France the proportion of moulting Eurasian reed warblers was only 3.2% of all the captures (most probably these birds were completing moult); therefore the estimates seem to be realistic. It is worth nothing that stopover duration in this study significantly increased with the progress of season, whereas in the Eastern Baltic (Courish Spit) it did not change much, from 6.0 days in early and mid August to 8.1 in mid and late September (Chernetsov 1998b).

In northern Spain blackcaps stopped over on average for 9.6 days (SE = 0.6; Arizaga et al. 2008). Interestingly enough, in this study both the probability of stay (φ) and seniority (γ) were found to be related to arrival body mass. It resulted in a negative relationship between the predicted stopover duration and arrival body mass, with stopover length estimate varying between 3.6 and 13.6 days (for the fattest and the leanest individuals, respectively).

Our estimates of the mean stopover duration of European robins on the Courish Spit were much shorter, only 1.1 days after the first capture, or 2.3 days with the time before the first capture included (Titov and Chernetsov 1999). It should however be emphasized that these estimates average all birds, 'transients' and non-transients (see Sect. 2.4). The most adequate estimate of the stopover length of non-transients is the estimate for the birds initially captured during the pauses of migration, 5.7 days (Titov and Chernetsov 1999).

The mean stopover duration of bluethroats *Luscinia svecica* in northern Karelia was 4.24 days (95% confidence interval 3.73–4.77 days; Panov and Chernetsov 2010). Ca. 78% of adult bluethroats make a one-day stopover; the remaining 22% stop over for 4.24 days, like juveniles.

Stopover duration estimates of grasshopper warblers *Locustella naevia* in southern England made on the basis of multiple day constancy models (Morris et al. 2005a) varied between 1.35 (SD = 0.95) and 2.13 days (SD = 1.02) in different years. Interestingly, MSL estimates that took single captures into account were very close, from 1.18 (SD = 0.94) to 1.30 days (SD = 1.21; Bayly and Rumsey 2007).

Estimates of duration of autumn stopovers of songbirds on Appledore Island (Maine, US) based on multiple-day constancy models taking into account the time before the first capture varied in different years between 2.70 (SE = 0.56; n = 354) and 3.45 days (SE = 0.67; n = 189) in the red-eyed vireo *Vireo olivaceus*; between 2.92 (SE = 1.95; n = 141) and 3.81 days (SE = 2.62; n = 99) in the American redstart *Setophaga ruticilla*; between 4.84 (SE = 1.50; n = 153) and 6.84 days (SE = 8.54; n = 270) in the northern waterthrush *Parkesia noveboracensis* (Morris et al. 2005a).

2.2.3 Estimates Based on Resightings

Salewski et al. (2007) studied stopovers of songbird migrants during spring migration across the Sahara in Ouadâne, an oasis in central Mauritania. They not only recaptured stopover migrants after the initial ringing, but also searched for colour-ringed birds and thus resighted individually marked individuals. It generally resulted in more precise estimates of probabilities of stay and thus of stopover duration. The mean estimates of stopover duration were 7.55 and 9.49 days (two different years) for the subalpine warbler *Sylvia cantillans*; 4.02 and 7.43 days for the whitethroat *Sylvia communis*; 6.58 and 7.75 days for the orphean warbler *Sylvia hortensis*; 6.29 days for the willow warbler (calculated from the data presented in Salewski et al. 2007).

An interesting experimental approach to estimating stopover duration was used by Bayly (2006). This author recorded the presence of individually marked Eurasian reed warblers at a feeding table with mealworms during the postbreeding period and autumn migration in England. The birds were recorded without capture, thus the potential capture effect was avoided. Migrants that once started to use supplementary food, continued to do so. The author interpreted the disappearance of the birds as migratory departure. The duration of the refuelling period estimated by this method was on average 5.82 days (SE = 0.20; n = 38). The author excluded search/settling time from this period, i.e. time preceding efficient mass gain.

Trying to solve this problem, in another study this author estimated the total stopover time of sedge warblers by adding to the observed duration of visiting feeding tables the mean of two estimates of the duration of stay before the birds started visiting feeders (Bayly 2007). One estimate was based on the assumption that before appearing at the feeder the birds gained mass at the same rate as when using the feeder (which seems a rather unrealistic assumption—N.C.); the other, on the assumption that the FDR of these birds was the mean of the upper quartile of sedge warblers captured and recaptured in mist-nets that have not received supplementary feeding. This complex procedure based on many assumptions of doubtful validity was performed for the birds (their number is not reported) whose body mass at the moment of the first visit at the feeder exceeded the mean body mass of sedge

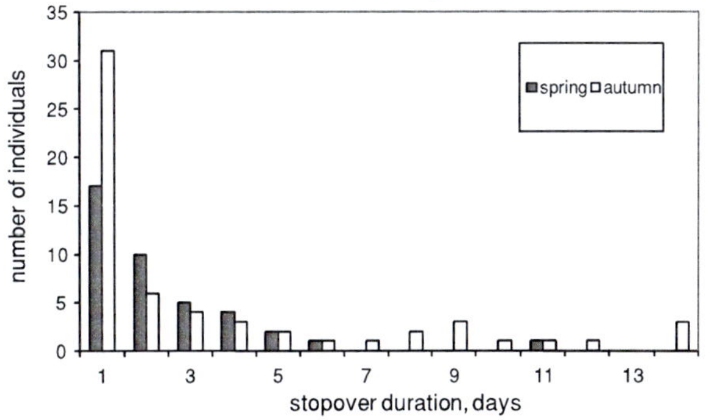

Fig. 2.2 Frequency distribution of stopover durations of radio-tagged European robins on the Courish Spit in spring (n = 40) and autumn (n = 59)

Table 2.3 Stopover duration of radio-tagged European robins on the Courish Spit on the Baltic coast in 2002–2003.

	Range of variation, days	Mean, days	Median, days	SE	Stopover > 2 days, %	n
Spring	1–12	2,4	2	0,31	30	40
Autumn	1–14	3,4	1	0,50	39	59

Only the birds tagged on the day of presumed arrival are included

warblers in that area by >1 g (Bayly 2007). The mean stopover duration estimated by this rather sophisticated method was 7.7 days (SE = 0.98; n = 20).

It should be noted that Bayly's experimental sedge warblers clearly formed two distinct clusters: those that stopped for 1–6 days (mean 3.8 days; SE = 0.47; n = 10) and gained mass up to 80% of the lean body mass at most; and those that stopped for 9–15 days (mean 11.6 days; SE = 0.69; n = 10) and gained 99–140% of the lean body mass. The sedge warblers from that latter group reached the body mass of more than 20 g and their FDR exceeded 13% per day which is at the very upper limit of the recorded FDR values (Bibby et al. 1976; Lindström 2003).

It should be also kept in mind that both Eurasian reed warblers (Bayly 2006) and sedge warblers (Bayly 2007) received supplementary food that allowed them to reach very high fuel deposition rates and may have very seriously influenced their stopover duration (see Chap. 4). These results might with some caution be used for studying the functional relationships between the stopover parameters, but should not be treated as the realistic estimates of stopover durations in the wild.

2.3 Estimates Based on Radio-Tagging

Radio-tagged European robins on the Courish Spit stopped for 1–11 days during spring migration, on average for 2.4 days (Fig. 2.2, Table 2.3). One bird that was not tagged on the first day upon arrival and because of that was not included into the analysis, stopped over for 12 days. Probably, 11–12 days is close to the upper limit of duration of spring stopovers of European robins.

In autumn, the stopover duration of European robins at the same site was 1–14 days, on average 3.4 days (Fig. 2.2, Table 2.3). The duration of migratory stopovers was not significantly different between the seasons (Mann–Whitney test, $z = 0.11$; $p = 0.92$). The autumn migratory stopovers of European robins on the Courish Spit were significantly shorter than in Mettnau on Lake Constance (SW Germany; 6.7 days, $SE = 1.04$; $n = 19$; Lajda 2001; Mann–Whitney test, $z = 2.79$; $p = 0.003$). Estimates from Mettnau are also based on radio-tagged birds, but in that study stopover duration might be biased high. Like in our study, wintering birds that have not left the study area before the end of study were excluded. However, the author did not follow nocturnal take-offs directly (Lajda 2001), therefore it cannot be claimed that all the individuals were indeed transient migrants at stopover. Some of them could have been wintering birds that changed their home range and were lost.

It is worth noting that the mean stopover duration estimated from telemetry data is very close to the estimate based on retrapping previously ringed birds in high mist-nets at the same site (3.3 days, calculated from the data in Bulyuk and Tsvey 2006).

We also radio tracked sedge warblers and pied flycatchers at stopover on the Courish Spit (Chernetsov et al. 2004). Of the 12 tagged sedge warblers seven stopped for one day, three for two and two for three days (mean 1.6 days, $SE = 0.23$). As pied flycatchers move at stopover significantly wider than sedge warblers or European robins (Chernetsov et al. 2004; Chernetsov 2005), it appeared possible to determine departure time in four individuals out of the seven tagged. The remaining three birds were lost during their diurnal movements, and we only have minimum estimates of their stopover duration, assuming that they took off on the night following their disappearance. Minimum estimates are that four birds made a one-day stopover, and two, three and four day stopovers were made by one individual each (mean 1.9 days, $SE = 0.46$).

Of 24 summer tanagers *Piranga rubra* radio-tagged during spring migratory stopovers immediately after crossing the Gulf of Mexico, 19 departed on the first night following arrival (i.e. made one-day stopovers), and five stopped over for more than one day (Moore and Aborn 1996). The exact stopover duration of these five birds remained unknown. It should be emphasized that even though summer tanagers were captured immediately after crossing the ecological barrier of at least 1,000 km of open water, 79% of them resumed migration on the first night upon arrival.

In a radio-tagging study of *Catharus* thrush stopovers at Long Point Bird Observatory on Lake Erie, Ontario, Canada, Swainson's thrushes *Catharus ustulatus* stopped over for an average of 3.4 days (up to 11.4 days) and hermit thrushes *C. guttatus* for an average of 8.9 days (up to 20.5 days; Mills et al. 2011).

Ovenbirds *Seiurus aurocapilla* radio-tracked in Brooklyn, New York City, in spring stopped over for 3.0 days on average (1–10 days, SE = 0.5, n = 27); 13 individuals made a one-day stopover (Seewagen et al. 2010). In autumn, they stopped over on average for 2.9 days (1–14 days, SE = 0.7, n = 22); 11 individuals departed in the first night upon arrival. All these values are very similar to the data from European robins on the Courish Spit.

A study of stopover duration of garden warblers on Crete in the eastern Mediterranean during autumn migration yielded very interesting results (Fransson et al. 2008). European robins, sedge warblers and pied flycatchers on the Courish Spit were studied during migration over optimal habitats; summer tanager stopped over immediately after crossing a barrier. Fransson et al. (2008) studied stopovers of migrants that were preparing for flights across half of the Mediterranean and the Sahara. The mean stopover duration was as long as 12.6 days (SE = 1.19; n = 20), and the authors do not rule out the possibility that birds tracked during 1–8 days were lost during local movements and did not embark on a migratory flight. If this is true, the stopover duration varied between 10 and 20 days, with the mean value of 15.3 days (SE = 0.63; n = 15). Thus, radio tracking data confirm the results of capture-mark-recapture studies in that before crossing major ecological barriers, passerine migrants make significantly longer stops than when migrating with a continuous stopover possibility.

2.4 Within-Species Variance in Stopover Duration: 'Transients' and 'Non-Transients'

Generally, stopover durations of 1–15 days are typical of small passerine nocturnal migrants that fly over ecologically suitable areas in Eurasia and North America with a continuous stopover opportunity. However, it is not always useful to estimate the *mean* duration of the stay of *all* migrants that make a stopover at a particular site. Not infrequently most birds stop for one day only. If a model with a time-since-marking-dependent probability of stay has the greatest support, some birds are transients with the probability of $\tau = 1 - \varphi_1/\varphi_2$. It was the case in Dzhanybek in western Kazakhstan (Chernetsov et al. 2007), in the Sahara (Salewski et al. 2007; Salewski and Schaub 2007), and, as our original data show, on the Courish Spit in the Eastern Baltic. During spring migration, in blackcaps the model that fitted the data best was φ (a2), p (.), i.e. with constant capture probability and time-since-marking-dependent probability of stay ($\varphi_1 = 0.146$; $\varphi_2 = 0.769$; $\tau = 0.81$). In whitethroats the best model was φ (a3), p (a2), i.e. with two time-since-marking classes of capture probability and three classes of probability of stay ($\varphi_1 = 0.175$; $\varphi_2 = 0.399$; $\varphi_3 = 0.838$; $\tau = 0.79$). In lesser whitethroats *Sylvia curruca*, a similar model received greatest support: $\varphi_1 = 0.161$; $\varphi_2 = 0.411$; $\varphi_3 = 0.855$; $\tau = 0.81$). In such situations it makes sense to estimate and to report the proportion of transients and stopover duration of non-transients. For example, in blackcaps on the Courish Spit the proportion of transients was 81%, the mean estimate of the stopover duration of non-transients was 3.81 days (–1/ln 0.769).

It has been suggested that all the birds that land after a migratory flight belong to one of the two groups: flyers, or transients, that are highly motivated to continue migration and therefore stop over for one or two days and feeders, or non-transients that mainly stop to refuel and thus intend to make a longer stopover (Rappole and Warner 1976). Stopover behaviour of the latter group is aimed at efficient foraging and refuelling which results in their search for optimal habitats and often competition-based interactions with conspecifics and heterospecifics (Rappole and Warner 1976; Dierschke and Delingat 2001). Usually or at least very frequently songbirds make several nocturnal migratory flights in a row (Fransson 1995; Bolshakov et al. 2003a, b; Hall-Karlsson and Fransson 2008). They make one-day stopovers between these flights, and a longer stopover when they complete such a series. As the aims of stopover in flyers and feeders are very different, their behaviour (foraging, spatial, and territorial) also may vary. However, to what extent the decision to continue migration in the next night after arrival is governed by the endogenous urge to migrate (i.e. is made even before arrival), and to what extent it depends on stopover events (habitat quality, food abundance, competition etc.) is a matter of debate. At least in European robins departure decisions are apparently governed by different environmental factors in birds that take off after short stopovers (1–2 days) and after longer stays with refuelling (Bulyuk and Tsvey 2006; Tsvey et al. 2007; Bolshakov et al. 2007).

It should be also kept in mind that estimates of the proportion of transients based on capture-mark-recapture data are usually seriously biased. It may happen if the range of daytime movements of migrants sharply declines some time after arrival at stopover. In such a case the proportion of birds that were captured within 1–2 days after their arrival can settle within a home range near the trapping site but outside of it. Some species do it regularly, as shown by our captures 500–1,000 m from the stationary trapping site. We repeatedly recaptured European robins and other migrants ringed at our trapping station beyond its borders. This is further supported by tracking of radio-tagged pied flycatchers (Chernetsov et al. 2004) and European robins (Chernetsov and Mukhin 2006). The proportion of radio-tagged European robins that departed from the study area on the Courish Spit on the first night upon arrival was much lower than the 87% suggested by capture-mark-recapture: 31 out of 59, i.e. 53% in autumn. In spring the proportion of one-day stopovers was even lower, 17 out of 40 (42.5%, Fig. 2.2). However, it is worth noting again that after crossing the Gulf of Mexico 79% of radio-tracked summer tanagers continued nocturnal migration on the first night after their arrival (Moore and Aborn 1996).

2.5 Concluding Remarks

Duration of migratory stopovers of songbirds usually varies between 1 and 15 days. Sometimes, especially before and just after crossing large ecological barriers (large water bodies, deserts) stopovers may be longer and reach 20–25 days, occasionally even longer. Significant proportions of migrants stop over for one day only and continue migration on the first night after arrival. When studying stopover behaviour by stochastic capture-mark-recapture models, it is most useful to estimate the proportion of 'transients' (migrants that make one-day stopovers) and the mean stopover duration of non-transients. However, it should be kept in mind that capture-mark-recapture models tend to overestimate the number of transients, whereas radio-tagging results in more realistic estimates.

Frequent occurrence of one-day stopovers means that many passerine nocturnal migrants make two or several nocturnal flights in succession, and then stop over for a longer period. It has been already suggested concerning several species of songbirds on the basis of analysing the distance of their flights from ringing data (Ellegren 1993; Fransson 1995) and fuel loads and potential flight range of birds captured during take-off and ceasing flight (Bolshakov et al. 2003a, b). The data on stopover durations support the existence of this pattern.

Such frequency distribution of stopover durations is reported from many regions. Except for areas before and immediately after ecological barriers, there are hardly any other unknown sites where nocturnal migrants gather for prolonged stopovers and refuelling. If most nocturnal migrants make several flights in a row, 'feeders' will always be a minority among the birds that arrive with a wave of migration. Certainly, it does not contradict the fact that some stopover sites are more suitable for refuelling than others, primarily because of varying habitat quality (Chap. 5; see also Ktitorov et al. 2008). Optimal migration theory predicts that stopover duration should be related to the fuel deposition rate (Alerstam and Lindström 1990; Lindström and Alerstam 1992; Hedenström and Alerstam 1997). Even though I criticise the optimal migration theory in Chap. 4, this statement seems to be correct. Fuel deposition rate is indeed one of the factors that govern stopover and departure decisions (Chap. 8).

References

Aborn DA, Moore FR (1997) Pattern of movement by summer tanagers (*Piranga rubra*) during migratory stopover: a telemetry study. Behaviour 134:1077–1100

Alerstam T, Lindström Å (1990) Optimal bird migration: the relative importance of time, energy, and safety. In: Gwinner E (ed) Bird migration. Springer, Berlin

Arizaga J, Barba E, Belda EJ (2008) Fuel management and stopover duration of blackcaps *Sylvia atricapilla* stopping over in northern Spain during autumn migration period. Bird Study 55:124–134

Bairlein F (1995) Manual of field methods, European-African songbird migration network. Institut für Vogelforschung, Wilhelmshaven

Bairlein F (1998) The European-African songbird migration network: new challenges for large-scale study of bird migration. Biol Cons Fauna 102:13–27

Balança G, Schaub M (2005) Post-breeding migration ecology of reed *Acrocephalus scirpaceus*, moustached *A. melanopogon* and Cetti's warbler *Cettia cetti* at a Mediterranean stopover site. Ardea 93:245–257

Bayly NJ (2006) Optimality in avian migratory fuelling behaviour: a study of a trans-Saharan migrant. Anim Behav 71:173–182

Bayly NJ (2007) Extreme fattening by sedge warblers, *Acrocephalus schoenobaenus*, is not triggered by food availability alone. Anim Behav 74:471–479

Bayly NJ, Rumsey SJR (2007) Grasshopper warbler *Locustella naevia* autumn migration—findings from a study in southeast Britain. Ring Migr 23:147–155

Bibby CJ, Green RE (1981) Autumn migration strategies of reed and sedge warblers. Ornis Scand 12:1–12

Bibby CJ, Green RE, Pepler GRM, Pepler PA (1976) Sedge warbler migration and reed aphids. Brit Birds 69:384–399

Blyumental TI, Gavrilov VM, Dolnik VR (1967) O prichinakh volnoobraznosti migratsii zyablika (On the causes of wave-like pattern of the chaffinch migration). Comm Baltic Comm Study Bird Migr 4:69–80

Bolshakov CV, Bulyuk VN, Sinelschikova A (2000) Study of nocturnal departures in small passerine migrants: retrapping of ringed birds in high mist-nets. Vogelwarte 40:250–258

Bolshakov C, Bulyuk V, Chernetsov N (2003a) Spring nocturnal migration of Reed Warblers *Acrocephalus scirpaceus*: departure, landing and body condition. Ibis 145:106–112

Bolshakov CV, Bulyuk VN, Mukhin A, Chernetsov N (2003b) Body mass and fat reserves of Sedge Warblers during vernal nocturnal migration: departure versus arrival. J Field Ornithol 74:81–89

Bolshakov CV, Chernetsov N, Mukhin A, Bulyuk VN, Kosarev V, Ktitorov P, Leoke D, Tsvey A (2007) Time of nocturnal departures in European robins, *Erithacus rubecula*, in relation to celestial cues, season, stopover duration and fat stores. Anim Behav 74:855–865

Bulyuk VN (2006) At what time of the day do passerine nocturnal migrants arrive at their breeding sites? Ardea 94:132–139

Bulyuk VN, Mukhin A, Fedorov VA, Tsvey A, Kishkinev D (2000) Juvenile dispersal in Reed Warblers *Acrocephalus scirpaceus* at night. Avian Ecol Behav 5:45–63

Bulyuk VN, Tsvey A (2006) Timing of nocturnal autumn migratory departures in juvenile European robins (*Erithacus rubecula*) and endogenous and external factors. J Ornithol 147:298–309

Burnham KP, Anderson DR (2002) Model selection and inference: a practical information-theoretic approach, 2nd edn. Springer, NY

Chernetsov N (1998a) On the direction of postnesting dispersal in the sedge warbler (*Acrocephalus schoenobaenus*) (Passeriformes, Sylviidae). Russ J Zool 2:626–628

Chernetsov N (1998b) Stopover length and weight change in juvenile Reed Warblers *Acrocephalus scirpaceus* in autumn in the Eastern Baltic. Avian Ecol Behav 1:68–75

Chernetsov N (2005) Spatial behavior of medium and long-distance migrants at stopovers studied by radio tracking. Annals N Y Acad Sci 1046:242–252

Chernetsov N (2010) Migratory stopovers of wrens *Troglodytes troglodytes* on the south-eastern Baltic coast. Avian Ecol Behav 17:13–22

Chernetsov N, Bulyuk VN, Ktitorov P (2007) Migratory stopovers of passerines in an oasis at the crossroad of the African and Indian flyways. Ring Migr 23:243–251

Chernetsov N, Mukhin A (2006) Spatial behaviour of European robins *Erithacus rubecula* during migratory stopovers: a telemetry study. Wilson J Ornithol 118:364–373

Chernetsov N, Mukhin A, Ktitorov P (2004) Contrasting spatial behaviour of two long-distance passerine migrants at spring stopovers. Avian Ecol Behav 12:53–61

Chernetsov N, Titov N (2000) Design of a trapping station for studying migratory stopovers by capture-mark-recapture analysis. Avian Ecol Behav 5:27–33

Cooch E, White E (2005) Program MARK: a gentle introduction, 5th edn. http://www.phidot.org/software/mark/docs/book/. Accessed 28 Nov 2010

Dierschke V, Delingat J (2001) Stopover behaviour and departure decision of northern wheatears, *Oenanthe oenanthe*, facing different onward non-stop flight distances. Behav Ecol Sociobiol 50:535–545

Dolnik VR (1975) Migratsionnoye sostoyanie ptits (Migratory disposition of birds). Nauka, Moscow

Efford M (2005) Migrating birds stop over longer than usually thought: comment. Ecology 86:3415–3418

Ellegren H (1993) Speed of migration and migratory flight lengths of passerine birds ringed during autumn migration in Sweden. Ornis Scand 24:220–228

Floerike C (1893) Vogelleben auf der Kurischen Nehrung. Ornithol Jahrb 4:1–11

Forstmeier W (2002) Benefits of early arrival at breeding grounds vary between males. J Anim Ecol 71:1–9

Fransson T (1995) Timing and speed of migration in North and West European populations of *Sylvia* warblers. J Avian Biol 26:39–48

Fransson T, Barboutis C, Mellroth R, Akriotis T (2008) When and where to fuel before crossing the Sahara desert—extended stopover and migratory fuelling in first-year garden warblers *Sylvia borin*. J Avian Biol 39:133–138

Gauthreaux SA Jr (1971) A radar and direct visual study of passerine spring migration in southern Louisiana. Auk 88:343–365

Grüll A, Zwicker E (1981) Nachbrutzeitliche Ortsveränderungen von Schilfrohrsänger (*Acrocephalus schoenobaenus*) und Teichrohrsänger (*A. scirpaceus*). Egretta 25:23–26

Hall-Karlsson KSS, Fransson T (2008) How far do birds fly during one migratory flight stage? Ring Migr 24:95–100

Hargrove JW, Borland CH (1994) Pooled population parameter estimates from mark-recapture data. Biometrics 50:1129–1141

Hedenström A, Alerstam T (1997) Optimum fuel loads in migratory birds: distinguishing between time and energy minimization. J Theor Biol 189:227–234

Herremans M (1990) Body-moult and migration overlap in reed warblers (*Acrocephalus scirpaceus*) trapped during nocturnal migration. Gerfaut 80:149–158

Jenni L, Winkler R (1994) Moult and ageing of European passerines. Academic Press, London

Kaiser A (1993) Rast- und Durchzugstrategien mitteleuropäischer Singvögel, Analysen von Fang- und Wiederfangdaten von Fangstationen zur Beschreibung der Ökophysiologie und des Verhaltens rastender Populationen. Doktorarbeit, Universität Konstanz

Kaiser A (1995) Estimating turnover, movements and capture parameters of resting passerines in standardized capture-recapture studies. J Appl Stat 22:1039–1047

Kaiser A (1999) Stopover strategies in birds: a review of methods for estimating stopover length. Bird Study 46(Suppl):S299–S308

Kokko H (1999) Competition for early arrival in migratory birds. J Anim Ecol 68:940–950

Ktitorov P, Bairlein F, Dubinin M (2008) The importance of landscape context for songbirds on migration: body mass gain is related to habitat cover. Landsc Ecol 23:169–179

Lajda M (2001) Telemetrische Untersuchung zum Rastverhalten des Rotkehlchens (*Erithacus rubecula*) in Südwestdeutschland während des Herbstzuges. Diplomarbeit, Universität Zürich

Lavée D, Safriel UN, Meilijson I (1991) For how long do trans-Saharan migrants stop over at an oasis? Ornis Scand 22:33–44

Lebreton J-D (2001) The use of bird rings in the study of survival. Ardea 89:85–100

Lebreton J-D, Burnham KP, Clobert J, Anderson DR (1992) Modeling survival and testing biological hypotheses using marked animals: a unified approach with case studies. Ecol Monogr 62:67–118

Lindström Å (2003) Fuel deposition rates in migrating birds: causes, constraints and consequences. In: Berthold P, Gwinner E, Sonnenschein E (eds) Avian migration. Springer, Heidelberg
Lindström Å, Alerstam T (1992) Optimal fat loads in migrating birds: a test of the time-minimization hypothesis. Am Nat 140:477–491
Mills AM, Thurber BJ, Mackenzie SA, Taylor PD (2011) Passerines use nocturnal flight for landscape-scale movements during migratory stopover. Condor 113:597–607
Moore FR, Aborn DA (1996) Time of departure by summer tanagers (*Piranga rubra*) from a stopover site following spring trans-Gulf migration. Auk 113:949–952
Morris SR, Liebner DA, Larracuente AM, Escamilla EM, Sheets HD (2005a) Multiple-day constancy as an alternative to pooling for estimating mark-recapture stopover length in Nearctic-Neotropical migrant landbirds. Auk 122:319–328
Morris SR, Turner EM, Liebner DA, Larracuente AM, Sheets HD (2005b) Problems associated with pooling mark-recapture data prior to estimating stopover length for migratory passerines. In: Ralph CJ, Rich TD (eds) Bird conservation implementation and integration in the Americas: Proceedings of the 3rd International Partners in Flight conference 2002, USDA Forest Service General Technical Report PSW-191, Albany
Mukhin A (2004) Night movements of young reed warblers (*Acrocephalus scirpaceus*) in summer: is it postfledging dispersal? Auk 121:203–209
Mukhin A, Chernetsov N, Kishkinev D (2008) Acoustic information as a distant cue for habitat recognition by nocturnally migrating passerines during landfall. Behav Ecol 19:716–723
Mukhin A, Kosarev V, Ktitorov P (2005) Nocturnal life of young songbirds well before migration. Proc R Soc B 272:1535–1539
Ottosson U, Waldenström J, Hjort C, McGregor R (2005) Garden Warbler *Sylvia borin* migration in sub-Saharan West Africa: phenology and body mass changes. Ibis 147:150–157
Panov IN, Chernetsov NS (2010) Migratsionnaya strategiya varakushki (*Luscinia svecica*) v Vostochnoy Fennoskandii. Soobschenie 1: Osnovnye parametry migratsionnykh ostanovok (Migratory strategy of bluethroats, *Luscinia svecica*, in Eastern Fennoscandia. Part 1: Main stopover parameters). Proc Zool Inst 314:93–104
Paradis E, Baillie SR, Sutherland WJ, Gregory RD (1998) Patterns of natal and breeding dispersal in birds. J Anim Ecol 67:518–536
Pradel R (1996) Utilization of capture–mark–recapture for the study of recruitment and population growth rate. Biometrics 52:703–709
Pradel R, Hines JE, Lebreton J-D, Nichols JD (1997) Capture-recapture survival models taking account of transients. Biometrics 53:60–72
Pradel R, Schaub M, Jenni L, Lebreton J-D (2005) Migrating birds stop over longer than usually thought: reply. Ecology 86:3418–3419
Rappole JH, Warner DW (1976) Relationships between behavior, physiology and weather in avian transients at a migration stopover site. Oecologia 26:193–212
Rguibi-Idrissi H, Julliard R, Bairlein F (2003) Variation in the stopover duration of reed warblers *Acrocephalus scirpaceus* in Morocco: effects of season, age and site. Ibis 145:650–656
Salewski V, Schaub M (2007) Stopover duration of Palearctic passerine migrants in the western Sahara—independent of fat stores? Ibis 149:223–236
Salewski V, Thoma M, Schaub M (2007) Stopover of migrating birds: simultaneous analysis of different marking methods enhances the power of capture-recapture analysis. J Ornithol 148:29–37
Schaub M, Jenni L (2000) Body-mass of six long-distance migrant passerine species along the autumn migration route. J Ornithol 141:441–460
Schaub M, Jenni L (2001a) Stopover durations of three warbler species along their autumn migration route. Oecologia 128:217–227
Schaub M, Jenni L (2001b) Variation of fuelling rates among sites, days and individuals in migrating passerine birds. Funct Ecol 15:584–594
Schaub M, Pradel R, Jenni L, Lebreton J-D (2001) Migrating birds stop over longer than usually thought: an improved capture-recapture analysis. Ecology 82:852–859

Schmaljohann H, Becker PJJ, Karaardic H, Liechti F, Neaf-Daenzer B, Grande C (2011) Nocturnal exploratory flights, departure time, and direction in a migratory songbird. J Ornithol 152:439–452

Seber GAF (1982) The estimation of animal abundance and related parameters, 2nd edn. Macmillan, NY

Seewagen CL, Slayton EJ, Guglielmo CG (2010) Passerine migrant stopover duration and spatial behaviour at an urban stopover site. Acta Oecol 36:484–492

Titov N (1999a) Individual home ranges of Robins *Erithacus rubecula* at stopovers during autumn migration. Vogelwelt 120:237–242

Titov N (1999b) Home ranges in two passerine nocturnal migrants at a stopover site in autumn. Avian Ecol Behav 3:69–78

Titov NV, Chernetsov NS (1999) Stokhasticheskie modeli kak novy metod otsenki prodolzhitelnosti migratsionnykh ostanovok ptits (Stochastic models as a new method for estimating length of migratory stopovers in birds). Uspekhi Sovremennoi Biologii 119:396–403

Tsvey A, Bulyuk VN, Kosarev V (2007) Influence of energy condition and weather on departures of first-year European robins, *Erithacus rubecula*, from an autumn migratory stopover site. Behav Ecol Sociobiol 61:1665–1674

Verkuil YI, Wijmenga JJ, Hooijmeijer JCEW, Piersma T (2010) Spring migration of Ruffs *Philomachus pugnax* in Fryslân: estimates of staging duration using resighting data. Ardea 98:21–33

Vysotsky VG (1998) Prostoy sposob rasscheta istinnoy sredney prodolzhitelnosti ostanovki u migriruyuschikh ptits (A simple method of estimating the true mean stopover duration in migrating birds). Russ J Ornithol 52:3–19

Wang Y, Moore FR (1993) Relation between migratory activity and energetic condition among thrushes (Turdinae) following passage across the Gulf of Mexico. Condor 95:934–943

Wang Y, Moore FR (1997) Spring stopover on intercontinental migratory thrushes along the northern coast of the Gulf of Mexico. Auk 114:263–278

Williams BK, Nichols JD, Conroy MJ (2002) Analysis and management of animal populations. Academic Press, San Diego

Yosef R, Wineman A (2010) Differential stopover of blackcap (*Sylvia atricapilla*) by sex and age at Eilat, Israel. Naturwissenschaften 74:360–367

Chapter 3
Fuel Deposition Rate and Energy Efficiency of Stopovers

Abstract In this chapter I review the methods of estimating the fuel deposition rate (FDR) of stopover migrants. These methods are based either on body mass change in the birds captured at least twice, or on body mass relationship with the time of day in single captures, or on analysis of metabolites in blood plasma. There is no perfect method; every approach has its benefits and pitfalls. The empirical values of the FDR reported in the literature are reviewed, and various factors that influence the FDR during migratory stopovers are discussed. Very often the FDR does not remain constant throughout stopover: it is low or even negative during 1–2 days after arrival but subsequently increases. It may also drop again towards the end of stopover. The ecological and behavioural causes and implications are discussed.

3.1 Energy Stores of Migrants

As mentioned in the Introduction, energy stores of migrating birds include not only adipose tissue but also some proteins (Piersma 1990; Lindström and Piersma 1993). In the thrush nightingale *Luscinia luscinia* fat made up 80–82% of stores, water, 12–14% and protein, 5–6% in mass (Klaassen and Biebach 1994; Klaassen et al. 1997). However, it should be mentioned that in first-autumn sedge warblers fat constituted 100% of stores above the lean body mass of 11.55 g (n = 20; Baggot 1986). It is generally assumed that birds extract ca. 5–10% of energy for the sustained migratory flight from protein (Jenni and Jenni-Eiermann 1998). It seems to be a small proportion, but the point is that energy density of wet protein is eight times lower than that of adipose tissue (Jenni and Jenni-Eiermann 1998). If this proportion is correct, protein should make up between 30% (if is the source of 5% of energy) and 47% (the 10% of energy case) of fuel stores in mass. Thus, it may be that in reality, a smaller proportion of energy is extracted from protein.

N. Chernetsov, *Passerine Migration*, DOI: 10.1007/978-3-642-29020-6_3,

Apparently the amount of energy derived from protein varies between the species of songbirds.

It has been suggested that long-distance migrants rely on protein to a smaller extent than short- to medium-distance migrants do (Jenni-Eiermann and Jenni 1991) and that purely insectivorous passerines burn more protein than species that consume large quantities of fruit rich in carbohydrates (Gannes 2001). It has also been reported that elevated blood concentrations of urea, indicative of increased protein oxidation, were typical of emaciated songbirds captured in the Sahara during autumn passage, i.e. that burning protein might be the last resort of individuals that have run out of lipid fuel (Bairlein and Totzke 1992).

Anyway, the potential flight range is greatly dependent on the proportion of protein in fuel stores (Jenni-Eiermann and Jenni 2003). Unlike lipids or carbohydrates, protein is stored as functioning tissue, therefore migrating birds that extract energy by burning protein have to get it from the breast muscle, gut, leg muscles etc. (Schwilch et al. 2002; Bauchinger and Biebach 2005).

3.2 Methods of Estimating Fuel Deposition Rate

3.2.1 FDR Estimates From Recaptures

As is the case with the stopover duration, estimating FDR has several methodological issues. There are two main methods to estimate it: on the basis of the mass change in the birds captured at least twice, and based on the mean mass change of samples of the birds captured during the day.

In the former case the analysis is aimed at estimating the daily FDR, measured in g day^{-1}. One of the difficulties is that if captures are made during the whole day, the body mass of individuals generally increases towards dusk as compared with dawn, because they forage (Klein et al. 1971; Dolnik 1975). The time of the first and the last capture seriously influence the body mass measurements. Attempts have been made to circumvent this problem by restricting trapping (or analysis) to morning hours. Unfortunately, this solution is not a good one, because mass is gained at the greatest rate before noon (when the morning activity and foraging peak takes place), and varies to a smaller degree in the afternoon [our data, see also Bardin (1998) on winter fuel stores in great tits *Parus major*].

Sometimes body mass is corrected for capture time. Body mass is standardised, usually by linear regression of body mass on time of the day, and then body mass change between the first and the last capture is calculated (Moore and Kerlinger 1987; Moore and Wang 1991; Chernetsov 1998).

The most correct method of estimating FDR from recaptures is based on multiple regression models (Schaub and Jenni 2000). The idea is that by stepwise multiple regression two coefficients are estimated, the mass gain rate during the day (*a*) and fuel deposited rate (*b*) after correcting for mass loss during the night

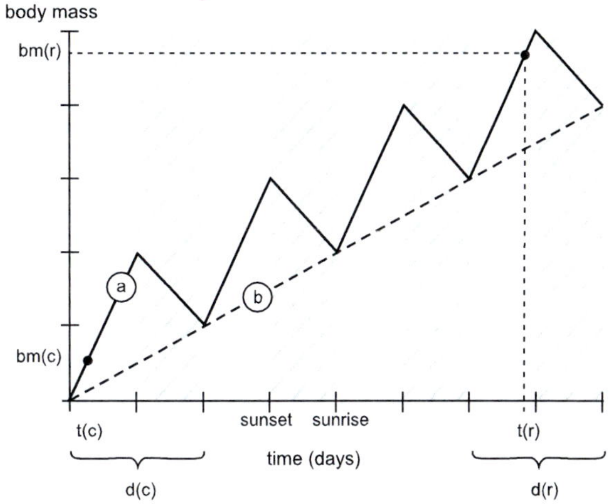

Fig. 3.1 A model of the body mass development of a migrant at stopover over several days (nights are shaded; altered from Schaub and Jenni 2000a). Explanations in the text

(Fig. 3.1). The initial capture and marking date [d(c)], time [t(c)] and body mass [bm(c)] are known for each individual. The respective parameters are also known for the recaptures [d(r), t(r), bm(r)]. This makes it possible to calculate body mass change during stopover from the equation:

$$\mathrm{bm(r)} - \mathrm{bm(c)} = a\,[\mathrm{daylength} - \mathrm{t(c)}] + b\,(\mathrm{d(r)} - \mathrm{d(c)} - a\,[\mathrm{daylength} - \mathrm{t(r)}], \tag{3.1}$$

which is equivalent to

$$\begin{aligned} \mathrm{bm(r)} - \mathrm{bm(c)} &= a\,[\mathrm{t(r)} - \mathrm{t(c)}] + b\,[\mathrm{d(r)} - \mathrm{d(c)}], \text{ or} \\ \mathrm{bm(r)} - \mathrm{bm(c)} &= a\,\Delta\mathrm{time} + b\,\Delta\mathrm{date} \end{aligned} \tag{3.2}$$

Thus, the dependent variable is mass change between the last and the first capture, and the independent variables (predictors) are the difference between the time of the last and the first capture (in hours) and the number of the days elapsed. When Δtime = 1 h, a is hourly mass change rate. When Δdate = 1 day, b is body mass change per day, i.e. FDR in g day^{-1}.

This method assumes that the body mass of stopover migrants linearly increases during the daytime and linearly decreases during the night (Fig. 3.1). In reality, deviations from the linear relationship between body mass and time of the day are usually small and not statistically significant (Schaub and Jenni 2000; Chernetsov and Titov 2001). An important advantage of this method is that it allows inclusion of additional parameters into the model and testing their significance. In the model, intercept is chosen to be equal to zero, as with zero time change mass change is obviously zero.

The disadvantage of this method is its dependence on recaptures. If the recapture probability varies between different cohorts of birds (e.g. because of their different mobility), artefacts may arise. Let us assume that all birds of some species intend to depart from the stopover site when they reach the same optimal departure body mass (or fuel load) and that both fat and lean individuals have the same mean FDR. Birds initially captured with a low body mass need relatively much time to reach the threshold body mass, during which individuals with both high and low FDR can be captured. Thus, for the initially lean birds we obtain a reliable mean FDR estimate. However, individuals with initially high body mass (close to the threshold) that gain mass slowly have a high probability of capture. But initially fat birds with high FDR quickly reach the threshold body mass and have a low probability of recapture during their brief stopover period. Therefore the mean FDR of fat birds is underestimated, and the overall mean FDR estimate is biased low, too. Moreover, it appears that FDR is inversely related to the initial body mass which is not actually true. Indeed, the inverse relationships between FDR and initial body mass (or fuel load) have been reported many times (Loria and Moore 1990; Fransson 1998; Schaub and Jenni 2000; Chernetsov 2001; Chernetsov and Titov 2001). It is impossible to decide whether this effect is an artefact on the basis of recaptures.

Last but not least, the estimates of FDR by captures and recaptures is by definition only possible for individuals that were captured at least twice. The assumption that these birds are a representative sample of all migrants at stopover is at least not self-evident and in most cases probably incorrect (Chap. 6).

3.2.2 FDR Estimates From Mass Change of First Captures

Another method of estimating FDR of migrants at stopovers is based on analysing the hourly trend of the mean mass of the first captures (Yablonkevich and Shapoval 1987; Winker et al. 1992; Dunn 2000). The resulting estimates are then not the daily, but the hourly mass change rate. Instead of the body mass change, the condition index change may be used, i.e. the change in size-corrected body mass (Labocha and Hayes 2012). As the proxy of structural size, wing length (Winker 1995; Dunn 2002; Ktitorov et al. 2008; Johnson and Winker 2008; Seewagen and Slayton 2008), or wing length raised to the third power (Dunn 2001) may be used. If the values obtained from different species are compared, FDR is expressed as percentage of the lean body mass (Lindström 1991, 2003). Hourly mass change rate may be compared with hourly nocturnal loss rate which is sometimes assumed to be 4.5% of lean body mass (Winker et al. 1992). However, estimates of nocturnal body mass loss rate based on the allometric equations may be more accurate (Dunn 2001). As an example of field data on body mass loss the data of Zimin (2003) on European robins may be given: during autumn migration the birds were losing on average 1.42 g per night (9.5%) under outdoor temperatures, with the average hourly rate of 1.2%. In chaffinches *Fringilla coelebs* in migratory disposition under outdoor temperature nocturnal mass loss was 0.103 g h^{-1} (SE = 0.011; Dolnik and Gavrilov 1973a), i.e. 0.5% per hour assuming the lean body mass of 20.0 g (Dolnik and Gavrilov 1973b). The daily mass gain should at least balance the nocturnal mass loss; otherwise the migrant will be losing mass instead of gaining it.

The estimates obtained by this method may be mainly used for comparing FDR values between different sites (Dunn 2002; Ktitorov et al. 2008). This data is difficult to use for the absolute estimates of mass change rate across several days, because estimates of nocturnal mass loss from allometric equations are rather rough and do not include thermoregulation costs that may differ considerably with calendar dates, geographic regions or even individually (Zimin 2003).

An important advantage of estimating mass change rate from single captures is that this approach does not exclude individuals that are only trapped once. Such birds are a great majority in any trapping project. This approach may only be used if all migratory flights are only performed at night: all the birds must arrive before the trapping starts. Otherwise individuals that arrive later during the day may be leaner (having depleted their fuel stores) than those captured early in the morning. Such situations may occur on offshore islands or on the coast where the migrants arrive after crossing large water bodies (Moore and Aborn 1996). Under such conditions even species that are normally pure nocturnal migrants may be forced to continue their flights after sunrise.

3.2.3 FDR Estimates From Blood Metabolites

The estimates based on multiple captures are limited by the fact that to be included, any bird should be captured at least twice. The method based on the hourly change of the mean body mass makes a non-obvious assumption that the diel rhythm of activity is the same in stopover migrants with different fuel loads. It has been shown that at least in some songbird migrants this is not what happens (Yablonkevich et al. 1985; Brensing 1989; Titov 1999b).

An attempt to circumvent these problems has been made by estimating the fuel deposition rate in migrants captured only once from concentrations of certain metabolites in blood plasma (mainly triglycerides, free glycerol and β-hydroxybutyrate). Plasma levels of triglycerides are positively, and those of β-hydroxybutyrate negatively correlated with body mass changes during several hours preceding sampling (Jenni-Eiermann and Jenni 1994; Williams et al. 1999; Jenni and Schwilch 2001; Cerasale and Guglielmo 2006). Most triglycerides originate from diet, either directly, or through synthesis in the liver; therefore high triglyceride concentrations indicate lipid transport to the peripheral tissues, i.e. fat deposition. β-hydroxybutyrate is a ketone body synthesized from fatty free acids which replaces glucose under the negative energy balance. Its high concentration in the plasma indicates the catabolism of fat. Therefore, high concentrations of these products are indicative of lipogenesis or catabolism of lipids in the organism. The plasma concentration of the third important metabolite, free glycerol, increases during lipid catabolism (Stevens 2004).

The study of the fuel deposition rate by analysing plasma concentrations of certain metabolites that indicate processes of lipid synthesis or catabolism is a promising avenue of research. Its obvious benefit is that a bird needs to be captured only once, so that the handling effect during the previous capture is absent. On the other hand, the results of analysis make it possible to determine processes that prevailed during the several hours preceding blood sampling. As FDR may seriously change during the stopover period (see below, Sect. 3.5), the data may not be representative of the whole duration of the stay. Moreover, the data obtained by this method allows comparisons between species, seasons, study sites (Schaub and Jenni 2001; Gannes 2001) or habitats (Leist 2007), but it is not straightforward to obtain absolute FDR values (in grams per hour or per day) from the plasma concentrations of metabolites. Attempted have been made to predict the body mass change rate from plasma metabolite concentrations in Eurasian reed warblers (Jenni and Schwilch 2001). However, at least 50% of body mass change variation remained unexplained, and it remained unclear how universal was the calibration equation obtained. It is not impossible that every songbird species will require its own species-specific calibration coefficient, which would make using this method for obtaining absolute estimates of FDR rather problematic.

3.3 Empirical FDR Values

3.3.1 The Mean and Maximum Values Observed

The FDR of songbird migrants during stopovers is usually ca. 1–3% of lean body mass per day (Lindström 2003). The median value for 31 species of passerine migrants was 2.4% (Biebach 1996). In small birds FDR (as the percentage of their lean body mass) was higher than in large ones. The maximum values of the mean FDR known for a species were related to the body mass as $FDR_{max} = 2.17\ m^{-0.34}$ ($r^2 = 0.54$; $p < 0.001$; 95% confidence interval of the exponential coefficient -0.44 to (-0.23); Lindström 2003). The highest FDR values reported in passerines were 12.4% of lean body mass per day in sedge warblers (Gladwin 1963), 12.5% in garden warblers and 13.0% in Eurasian reed warblers (N. Baccetti in Lindström 2003). It remains unclear how these values were obtained. They are probably not the net daily FDR (*b* coefficient in Fig. 3.1), but the mass increase during the day (*a* coefficient). Similarly high values (11.2–13.6% of lean body mass per day) were obtained for the grey catbird *Dumetella carolinensis*, northern waterthrush, red-eyed vireo, and several *Empidonax* flycatchers during the autumn passage in southern Belize (Johnson and Winker 2008). However, these values may also be biased high due to the underestimated values of nocturnal body mass loss in that study.

The physiological limit of mass accumulation rate when food is provided *ad libitum* is even higher: some sedge warblers increased their body mass by more than 20% of their lean body mass per day in a field experiment with supplementary food provision (Bayly 2006). Apparently, the migrating passerines never or nearly never reach such FDR values in the wild. Most field data yield much lower values. It should be however kept in mind that most empirical values are more or less biased low.

Data for 48 species of songbirds captured on the shore of Lake Erie in North America based on the increase of the mean mass of the first captures during the day give the mean value of 0.50% of lean body mass per hour during spring passage and 0.61% in autumn (Dunn 2001). The data for 14 species at 15 sites in southern Canada yield 0.40% of lean body mass per hour in spring and 0.53% in autumn (Dunn 2002). These values are very close to FDR values of European robins based on recaptures (see below).

3.3.2 *Case Study: The European Robin on the Courish Spit*

A detailed analysis of FDR at stopover and factors that influence it was made by Tsvey (2008) in a case study of European robins on the Courish Spit in the Eastern Baltic. In order to estimate the FDR during the day (body mass increase rate during the day, g h^{-1}) and the net FDR during the stopover (g day^{-1}) the author selected 895 capture histories of European robins (177 in spring and 718 in autumn) trapped in 1994–2003 on Cape Rossitten on the Courish Spit. In spring, captures occurred on 27 March–15 May and involved 33 adults and 144 second-year birds. In autumn, 40 adults and 678 hatching-year birds were captured between 1 September and 6 November. All the adults had their post-breeding moult completed; 55 first-autumn robins were at final stages of juvenile moult.

The capture histories selected for analysis met the following four conditions. First, only those capture histories of birds were included which had been initially captured during the first or the second day of a wave of arrivals. The first capture at the beginning of a wave strongly suggested that the bird had been marked soon after arrival (Sect. 2.1, see also Titov and Chernetsov 1999b; Chernetsov and Titov 2000). The initial body mass of such birds is more or less representative of arrival fuel stores.

Second, only those cases were included in which the minimum stopover duration (time elapsed between the first and the last capture) did not exceed 12 days in spring and 17 days in autumn. These values were selected on the basis of frequency distribution of stopover durations of European robins when the capture histories met the first criterion, and on the basis of radio-tracking data (Sect. 2.3) and of recaptures in high mist-nets.

Third, to reduce the bias caused by handling effect, only those cases were considered when at least two days elapsed between the initial marking and the first recapture. Finally, in the cases of multiple recaptures only the body masses at the first and the last capture were included. If multiple captures occurred during the last and penultimate day of proven stay, only the first of these captures was considered. For example, if a robin was captured three times on 18 September and once more on 19 September, the first capture on 18 September was analysed. This was done to minimise the possible effect of handling stress.

By following these selection criteria, Tsvey (2008) analysed the data that referred to birds that stopped over for three and more days, i.e. were 'feeders', not 'flyers' sensu Rappole and Warner (1976); and were captured soon after arrival. The author included eight potential predictors of the rate of body mass change into the multiple regression model. These predictors were (1) season (spring vs. autumn); (2) progress of season; (3) age of birds (first- or second-year vs. adults); (4) condition index at the first capture, calculated as CI = body mass · wing length$^{-0.852}$, where body mass is standardised for 8 a.m., and exponent follows Titov and Chernetsov (1999a); (5) wing length as size proxy; (6) number of European robins at the study site during stopover of each individual birds as a proxy of density of conspecifics; (7) air temperature at midday; (8) year.

The best model selected by backward stepwise elimination and including both spring and autumn data explained 26% of FDR variance ($r^2 = 0.26$, $F_{3,891} = 103.1$, $p < 0.0001$, $n = 895$) and resulted in the estimate of 0.084 g h^{-1}. It makes 0.57% of lean body mass of an average European robin with a wing length of 72 mm. It is noteworthy that when large samples are analysed, the estimates obtained by different methods are very similar. The analysis of the first captures across time of the day on Cape Rossitten in 1994–2001 suggested a body mass increase of 0.0804 g·h^{-1} (body mass = 15.15 + 0.0804 h, $r^2 = 0.072$, $n = 46{,}730$, $p < 0.01$), i.e. 0.5% per hour. Captures of European robins in elevated mist-nets after one-day stopovers yielded FDR estimate of 0.0823 g h^{-1} (SD = 0.04, n = 24; Bulyuk and Tsvey 2006). The latter value is based on two captures of individual birds.

The net FDR (coefficient *b*) after correction for the time of the day was 0.064 g day^{-1} for autumn migration (0.44% of lean body mass) and only 0.007 g day^{-1} in spring (Tsvey 2008). This means that in spring, the FDR of European robins on the Courish Spit was very close to zero. Estimates for the autumn passage are close to the figures reported from Helgoland in the North Sea, based on recaptures: 0.04–0.11 g day^{-1}, depending on the progress of season (Ottich and Dierschke 2003).

During spring passage, both mass increase during the day and net energy efficiency of stopover were inversely related to the initial condition index (Table 3.1). Lean European robins (with the mean condition index of 0.385, i.e. mean body mass of 14.72 g) increased their body mass by 0.09 g h^{-1}, whereas fat robins (with the mean condition index of 0.427, i.e. mean body mass of 16.33 g), by 0.06 g h^{-1}. Lean individuals gained mass by 0.065 g day^{-1}, and their fat conspecifics lost mass by 0.11 g day^{-1}. In 2001 FDR was significantly (by 0.26 g day^{-1}) higher than in other years, and even fat European robins increased their body mass at stopovers on the Courish Spit (on average by 0.15 g day^{-1}).

In autumn, body mass gain during the day was positively related to the progress of the season (Table 3.2), i.e. foraging efficiency increased from 0.083 g h^{-1} on 1 September to 0.104 g h^{-1} on 31 October. The energy efficiency of stopovers was significantly related to their initial fuel stores and the number of birds present at stopover (a proxy of competition). In 1995 FDR was significantly lower than in the remaining years, other factors being held equal (Table 3.2). Lean European robins increased their body mass on average by 0.13 g day^{-1}, whereas the body mass of

Table 3.1 Multiple regression model explaining the rate of the body mass change in European robins on the Courish Spit during spring passage (from Tsvey 2008)

Parameter	Regression coefficient	SE	Significance
Δtime	0.370	0.134	$p < 0.0001$
Δtime·CI	−0.727	0.330	$p < 0.05$
Δdate	1.595	0.212	$p < 0.0001$
Δdate·CI	−3.988	0.527	$p < 0.0001$
Δdate·year 2001	0.255	0.079	$p < 0.001$

$R^2 = 0.46$; $F_{5,172} = 28.9$, $p < 0.0001$, $n = 177$. Δtime is the hourly mass change rate; Δdate is the daily mass change rate taking account of nocturnal mass loss; CI is condition index (size-corrected body mass) at the first capture

Table 3.2 Multiple regression model explaining the rate of body mass change in European robins on the Courish Spit during autumn passage (from Tsvey 2008)

Parameter	Regression coefficient	SE	Significance
Δtime·season progress	0.00034	0.00003	$p < 0.0001$
Δdate	1.277	0.100	$p < 0.0001$
Δdate·CI	−2.878	0.245	$p < 0.0001$
Δdate·number	−0.0003	0.00008	$p < 0.0001$
Δdate·year 1995	−0.0093	0.018	$p < 0.0001$

$R^2 = 0.41$; $F_{5,713} = 100.5$, $p < 0.0001$, $n = 718$. Δtime is the hourly mass change rate; Δdate is the daily mass change rate taking account of nocturnal mass loss; CI is condition index (size-corrected body mass) at the first capture; number is the mean daily trapping figure during the migratory stopover of a particular bird

fat birds practically did not change: their FDR was 0.011 g day^{-1} (the mean number of conspecifics at the stopover site was 122 individuals). In 1995, fat robins increased their body mass by 0.04 g day^{-1}, lean birds lost body mass by 0.08 g day^{-1}.

3.3.3 FDR Estimates From Blood Metabolites Analysis

A broad-scale field study of fuel deposition rate by analysing blood metabolites in the Eurasian reed warbler, sedge warbler, garden warbler and pied flycatcher was performed by Schaub and Jenni (2001). The authors reduced two parameters, plasma levels of triglycerides and β-hydroxybutyrate, to the single 'fattening index' by principal component analysis. They translated the values of this fattening index into body mass change from sunrise to sunset (i.e. coefficient *a*) for Eurasian reed warblers using the calibration equation developed by Jenni and Schwilch (2001). The results varied between 0.190 g over the daylight hours in Ebro delta (Spain) and 1.028 g in Oued Moulouya (Morocco, just before crossing the Sahara in autumn). These estimates seem to be the realistic ones.

3.4 Factors that Influence FDR

3.4.1 *Effect of Initial Fuel Stores*

The factor that nearly always significantly negatively influences FDR of migrants is their energy condition (condition index) at the first capture (Loria and Moore 1990; Fransson 1998; Schaub and Jenni 2000; Chernetsov 2001, 2003, 2010; Bayly and Rumsey 2007; Chernetsov et al. 2007; Tsvey 2008; Panov and Chernetsov 2010). It has already been mentioned (Sect. 3.2.1) that it may be an artefact. It is however worth noting that mobility of fat and lean European robins in the first two days of stopover on the Courish Spit did not differ (Sect. 6.5), so that the capture probability should not necessarily be fuel load-related. The capture probability of several species of long-distance migrants in an oasis in the western Sahara has also been shown to be independent of fuel load (Salewski and Schaub 2007). Under experimental conditions caged fat European robins lose body mass during the first week after capture, whereas lean birds start to gain body mass immediately after being taken into captivity (Tsvey 2008). Song thrushes *Turdus philomelos* do the same (Chernetsov, unpubl.). Therefore it cannot be ruled out that at least in some cases the inverse relationship between the initial body mass and FDR is not due to methodological issues but is a real phenomenon. Migrants with high fuel stores probably invest more time and effort into caring about their safety from predators (Fransson and Weber 1997) and spend less time in habitats rich in food but more open and thus providing less protection from predators (Moore 1994; Dierschke 2003; Sapir et al. 2004; Wang and Moore 2005).

Only a single study reported a positive relationship between current body mass and FDR in stopover migrants (Schaub and Jenni 2001). In this study FDR was estimated not from recaptures but from blood metabolite analysis (Sects. 3.2.3, 3.3.3). In three species out of four included in the analysis (in the Eurasian reed warbler, sedge and garden warblers) the lipid synthesis rate was positively correlated with body mass. This study was free from the potential methodological issues discussed in Sect. 3.2.1, when the FDR estimates of initially fat birds may be biased low. On the other hand, it cannot be ruled out that the amount of stored lipids also influences the level of triglycerides, free glycerol and β-hydroxybutyrate, apart from the rate of lipid catabolism.

3.4.2 *Progress of Season*

Another factor that is often included into multiple regression models as a significant predictor of FDR is progress of season. In the sedge warbler and Eurasian reed warbler across most of Europe (Schaub and Jenni 2000; Balança and Schaub 2005), in the blackcap on the Courish Spit (Chernetsov and Titov 2001), in the whitethroat in Sweden (Fransson 1998), in the European robin on the Courish Spit

(Tsvey 2008) and on Helgoland (Ottich and Dierschke 2003) the energy efficiency of stopovers increased towards the end of autumn passage. However, the FDR of garden warblers across Europe (Schaub and Jenni 2000) and bluethroats in northern Karelia (Panov and Chernetsov 2010) was not related to progress of season. Increasing FDR in the end of autumn allows migrants to deposit greater fuel stores even without increasing stopover duration. It makes it possible to make longer flights and thus to increase the overall speed of migration.

The adaptive value of high FDR at the end of the autumn migratory season is obvious. Delayed individuals need to leave quickly the areas where the weather and foraging opportunities rapidly deteriorate or at least become unpredictable. Indeed, the speed of migration is known to increase towards the end of the season (Ellegren 1993; Fransson 1995; Bensch and Nielsen 1999). Furthermore, it may be essential for a number of species to arrive early or at least not too late to wintering areas, just like it is important to arrive first in spring (Kokko 1999). It concerns species like the pied flycatcher or European robin that are territorial in winter quarters (von Stünzner-Karbe 1996; Cuadrado 1997; Salewski 1999; Tellería and Péres-Tris 2004) and like *Acrocephalus* warblers for whom early arrival at Africa may make it possible to undergo moult at the beginning of winter (Bensch et al. 1991).

It is less clear how the birds increase their fuel deposition rate at the end of the autumn migratory season. It is usually assumed that food availability for the insectivorous songbirds declines at the end of autumn. This is not so for the European robin, whose prey availability increases towards the end of the autumn migratory season in the Eastern Baltic, because hibernating invertebrates gather in the ground layer late in autumn and become more available for foraging robins (Titov 2000). It is worth noting that in this species the gain rate of body mass during the day (coefficient a) increases in late autumn, but the net FDR (coefficient b) does not (Table 3.2). How *Acrocephalus* warblers manage to increase their FDR in late autumn (Schaub and Jenni 2000, 2001; Balança and Schaub 2005), if they mainly consume highly seasonal plum aphids (Bibby and Green 1981; Chernetsov and Manukyan 1999a, b, 2000) and their prey availability does decline in late autumn, is more difficult to understand. Endogenously controlled increase in FDR towards the end of the autumn migratory period has been recorded in captive long-distance migrants under the laboratory conditions (Bairlein 2000). It seems that in many cases migrating songbirds do not gain mass at the maximum possible rate, and they are able to increase FDR when necessary, e.g. when delayed in northern areas in late autumn, or before crossing an ecological barrier.

3.4.3 Competition

An inverse relationship between the number of conspecifics present at the stopover site and FDR was found during autumn passage in goldcrests *Regulus regulus* in Sweden (Hansson and Pettersson 1989), in the Wilson's warbler *Wilsonia pusilla* in North America (Kelly et al. 2002) and in the European robin on the Courish Spit (Tsvey 2008). In spring, a similar relationship was reported from several species of long-distance passerine migrants after crossing the Gulf of Mexico (Moore and Wang 1991). It cannot be ruled out that lower FDR recorded in hatching-year individuals as compared to adults in a number of species, e.g. in the Savi's warbler *Locustella luscinioides* in Portugal (Neto et al. 2008) and in bluethroats in Karelia (Panov and Chernetsov 2010) was also due to the higher competitive ability of experienced migrants.

High concentrations of songbird migrants at stopovers may increase both direct and interference competition (Salewski et al. 2007a) and decrease the energy efficiency of stopovers (Moore and Wang 1991). In one case (in the blackcap on the Courish Spit, Chernetsov and Titov 2001) the number of birds at stopover was positively, not negatively, related to FDR. The main food of blackcaps during autumn passage in this area is common elder *Sambucus nigra* which may be superabundant at times. Probably the highest concentrations of blackcaps at stopovers on Cape Rossitten were recorded when the common elder was very abundant, thus enabling a high FDR. It is also worth noting that densities of blackcaps in this area are never as high as e.g. those of European robins or goldcrests. Very high numbers of the latter species may indeed cause strong competitive interactions and inhibit their fuelling rate.

3.4.4 Food Availability

Apparently, most frequently FDR of stopover migrants should depend on food availability but this relationship is only infrequently formally shown. To obtain reliable estimates of food abundance for insectivorous or omnivorous songbirds, which are mostly long-distance migrants, is methodologically rather challenging (Grosch 1995; Eggers 2000; Salewski 1999; Chernetsov and Manukyan 1999a, b). As an example of such study one may make a reference to the research done in northern Spain (Grandío 1998). At a site with a high abundance of plum aphids sedge warblers increased their body mass by 0.96 g day^{-1} (SD = 0.23, n = 12), whereas at a poor site FDR was just 0.46 g day^{-1} (SD = 0.31, n = 22, Mann–Whitney test: $p < 0.001$). These values correspond to 9.3 and 4.6% of lean body mass. It is important that at both sites, trapping was performed simultaneously and within a short time interval, 5–11 August 1995, making the results comparable (Grandío 1998).

Sometimes foraging manoeuvres are studied as a proxy to food choice and abundance (Chen et al. 2011). It should be however noted that foraging manoeuvres in birds were suggested to be highly stereotyped (Khlebosolov 1993, 1996, 2005) and not very suitable for characterising highly variable stopover behaviour.

It has been shown that diet influences the composition of blood metabolites and thus probably the composition of fuel stores in long-distance migrants (Gannes 2001). Fruit-eating species, like blackcaps, garden warblers and lesser whitethroats, catabolised less protein and more lipids during their spring passage in Israel.

Apart from predictors that may be accounted for, there are other factors that influence FDR but elude analysis. These factors e.g. provided high FDR in European robins on the Courish Spit in the spring of 2001 and low FDR in the autumn of 1995, after other predictors had been taken into account (Tables 3.1, 3.2). These factors also result in high variance of FDR values typical of all the study sites (Dunn 2001, 2002; Tables 3.1, 3.2). An important factor for insectivorous birds is probably air temperature, as shown by the analysis performed by Schaub and Jenni (2001) on the basis of data from Rybachy on the Courish Spit. For pied flycatchers that forage on highly mobile aerial prey air temperature was an important predictor of fattening index, whereas FDR garden warblers in autumn, when they are largely frugivorous, was independent of weather variables (Table 6 in Schaub and Jenni 2001).

3.5 Low Initial FDR: Artefact or Real Phenomenon?

Many authors have reported that body mass of songbird migrants drops after their arrival to stopover and it is not until several days after arrival that they start to gain mass (Szulc-Olech 1965; Pettersson 1983; Hansson and Pettersson 1989; Loria and Moore 1990; Mädlow 1997; Wang and Moore 1997; Titov 1999a; Yosef and Wineman 2010).

3.5.1 Methodological Issues

It is usually assumed that at least two methodological issues seriously limit the value of these observations: first, the first capture does not always occur immediately after arrival; and second, capture and associated loss of foraging time and handling stress may significantly reduce foraging efficiency and FDR (Schwilch and Jenni 2001).

The former issue is less severe than it is often assumed. The first capture may indeed take place several days after arrival, but in areas with a pronounced wave-like pattern of passage (like the Baltic coast, where most of the aforementioned European studies were performed, or the Gulf of Mexico coast, where North American studies were made) most birds are captured soon after arrival (Titov and Chernetsov 1999b). Thc authors write that '[A]ssuming capture probability to be independent of time since arrival, only a minority of first captures have… arrived the night before' (Schwilch and Jenni 2001). This is correct, but it is the assumption that capture probability is independent of time since arrival that is unjustified, because the mobility of most migrants (which governs capture probability) sharply drops after one or two days at stopover (Chap. 6).

The problem of the possible effect of handling stress was analysed by Schwilch and Jenni (2001) in much detail. These authors measured the plasma levels of triglycerides and β-hydroxybutyrate and showed that fattening rate did not differ between Eurasian reed warblers that had just arrived at the stopover site (tape-lured into the area) and those that had been there already during several days. The FDR of birds captured twice in one day was not lower than of those trapped only once. The authors claimed that since at their study area (Wauwiler Moos in Switzerland) mist-nets were checked every 20 min, and not every hour, which is the routine at most mist-netting sites (Bairlein 1995, 1998), mass change values, though negative, but not significantly different from zero, the trapping effect could be considered absent. It was different from two other Swiss sites (Portalban and Bolle di Magadino) where recapture data suggested mass loss on the day of capture (Schwilch and Jenni 2001).

I believe that the authors' claims are too far-fetched. The mean mass change on the day of capture in Wauwiler Moos was negative and, most importantly, significantly different from the positive value of mass change on the subsequent days (Fig. 3 in Schwilch and Jenni 2001). A similar pattern was observed at other trapping sites (Fig. 4 in Schwilch and Jenni 2001). Besides, the claim that the tape-lured Eurasian reed warblers had just arrived, whereas the individuals trapped without song playback had arrived several days before capture, is not undeniable (see above). Contrary to the authors' viewpoint, their data does not refute the opinion that migrants lose body mass on the first day upon arrival at stopover. In Louisiana, after crossing the Gulf of Mexico in spring, the probability of losing body mass in red-eyed vireos was not related to the number of recaptures (Loria and Moore 1990), i.e. handling stress is not the only reason for initial mass loss.

Recaptures and radio-tracking data suggest that mobility of many passerine migrants on the day of arrival to stopover and partly on the second day is significantly higher than on subsequent days (Chap. 6). It seems probable that during the first 1–2 days after arrival migrants select the optimal habitat and occupy temporary home ranges, i.e. perform search/settling activities, and only after they are settled, they start to refuel at the maximum possible rate. It happens even when no major barrier crossing is involved, when many passerines reduce their digestive organs during long-distance flights and need to build them up before they are able to forage at the maximum rate (Hume and Biebach 1996; Biebach 1998; Bauchinger and Biebach 1998; Karasov and Pinshow 1998). Low and sometimes even negative FDR on the first day of stopover in passerines is not an artefact caused by handling stress, but a rather widespread event.

3.5.2 Physiological and Ecological Constraints

Low initial FDR may be due to physiological constraints, especially after endurance flights when crossing ecological barriers (Karasov and Pinshow 1998; Biebach 1998; Bauchinger and Biebach 2001, 2005; Bauchinger et al. 2005), or due to difficulties that a migrant faces after completing flight and arriving at a novel area (see Chap. 5). The latter situation, i.e. ecological difficulties settled by behavioural adaptations, is probably the main reason for low FDR on the first day(s) after arrival when no barrier crossing is involved.

All data on digestive tract reduction that inhibits efficient foraging immediately after completing migratory flights, in songbirds (see the aforementioned references) and in waders (Piersma 1998; Piersma and Gill 1998; Piersma et al. 1999; Battley et al. 2000, 2001), refer to very long endurance flights. No data available suggests that similar changes may occur during migration with continuous stopover possibility, i.e. without very long flights. It seems that in such situations no significant reduction of the digestive tract occurs that might cause physiological difficulties for foraging at the maximum rate possible.

As body mass decline within 1–2 days upon arrival at stopover in songbirds is usually explained not by their physiology but by ecology (search/settling difficulties), in some optimal habitats this decline may not happen (Bairlein 1987; Moore and Kerlinger 1987; Carpenter et al. 1993). However, analysis of the literature suggests that body mass drop soon after arrival occurs more often than not.

A significant consequence of this is that FDR is generally not stable during the stopover period. At the beginning of a stopover FDR is small or even negative. Later on, if a migrant remains in the area, it increases. The optimal migration theory assumes that when the migrant reaches a certain threshold of fuel stores (optimal departure fuel load, with optimality defined by various parameters, and, no less importantly, by the researcher's many assumptions), it takes off, thus completing the stopover. This model however assumes that departure decision is made on the basis of energetic considerations only. In the wild, however, other factors are in play as well, in particular the weather (Liechti 1995, 2006; Liechti and Bruderer 1998; Åkesson and Hedenström 2000; Schaub et al. 2004; Tsvey et al. 2007). It is not known how large or small FDR is in a migrant that has reached the threshold fuel load but is delayed by the adverse weather. In some cases under such circumstances body mass may decrease again, which was found in whitethroats in Sweden (Fransson 1998) and in Eurasian reed warblers in Israel (Merom et al. 2000). The data on FDR of sedge warblers in England also supports the existence of such effect: both the mass increase rate during the day (coefficient a) and the net FDR (coefficient b) declined towards the day of departure (Bayly 2007). It cannot be ruled out that FDR declines in the last days preceding departure not only because of the adverse weather (which prevents departure), but also due to the elevated existence costs with high fuel stores (Klaassen and Lindström 1996).The latter authors assume that large fat stores have high metabolic costs of their maintenance, even though it should be kept in mind that adipose tissue is metabolically rather inert, therefore direct metabolic costs of its maintenance are not very high (Schmidt-Nielsen 1997). Moreover, fat individuals have higher costs of locomotion (especially of flight) and may have an inferior ability to escape predators (but see below, Sect. 4.4).

References

Åkesson S, Hedenström A (2000) Wind selectivity of migratory flight departures in birds. Behav Ecol Sociobiol 47:140–144

Baggot GK (1986) The fat contents and flight ranges of four warbler species on migration in North Wales. Ring Migr 7:25–36

Bairlein F (1987) The migratory strategy of the garden warbler: a survey of field and laboratory data. Ring Migr 8:59–72

Bairlein F (1995) Manual of field methods, European-African songbird migration network. Institut für Vogelforschung, Wilhelmshaven

Bairlein F (1998) The European-African songbird migration network: new challenges for large-scale study of bird migration. Biol Cons Fauna 102:13–27

Bairlein F (2000) Photoperiode und Nahrungsangebot beeinflüssen zugzeitliche Fettdeposition. Jahresbericht Institut für Vogelforschung 4:5

Bairlein F, Totzke U (1992) New aspects on migratory physiology of trans-Saharan passerine migrants. Ornis Scand 23:244–250

Balança G, Schaub M (2005) Post-breeding migration ecology of reed *Acrocephalus scirpaceus*, moustached *A. melanopogon* and Cetti's warbler *Cettia cetti* at a Mediterranean stopover site. Ardea 93:245–257

Bardin AV (1998) Opyt prilozheniya ridit-analyza k dannym vizualnoy otsenki zhirovykh reservov na primere analyza sutochnoy dinamiki zhirnosti bolshikh sinits *Parus major* v zimniy period (Application of ridit analysis to subcutaneous lipid data using daily fat deposit variation in the Great Tit *Parus major* during winter as an example). Russ J Ornithol 49:17–24

Battley PF, Piersma T, Dietz MW, Tang S, Dekinga A, Hulsman K (2000) Empirical evidence for differential organ reductions during trans-oceanic bird flight. Proc R Soc Lond B 267:191–195

Battley PF, Dietz MW, Piersma T, Dekinga A, Tang S, Hulsman K (2001) Is long-distance bird flight equivalent to a high-energy fast? Body composition changes in freely migrating and captive fasting knots. Physiol Biochem Zool 74:435–449

Bauchinger U, Biebach H (1998) The role of protein during migration in passerine birds. Biol Conserv Fauna 102:299–305

Bauchinger U, Biebach H (2001) Differential catabolism of muscle protein in garden warblers (*Sylvia borin*): flight and leg muscle act as a protein source during long-distance migration. J Comp Physiol B 171:293–301

Bauchinger U, Biebach H (2005) Phenotypic flexibility of skeletal muscles during long-distance migration of garden warblers: muscle changes are differentially related to body mass. Annals New York Acad Sci 1046:271–281

Bauchinger U, Wohlmann A, Biebach H (2005) Flexible remodeling of organ size during spring migration of the garden warbler (*Sylvia borin*). Zoology 108:97–106

Bayly NJ (2006) Optimality in avian migratory fuelling behaviour: a study of a trans-Saharan migrant. Anim Behav 71:173–182

Bayly NJ (2007) Extreme fattening by sedge warblers, *Acrocephalus schoenobaenus*, is not triggered by food availability alone. Anim Behav 74:471–479

Bayly NJ, Rumsey SJR (2007) Grasshopper warbler *Locustella naevia* autumn migration—findings from a study in southeast Britain. Ring Migr 23:147–155

Bensch S, Nielsen B (1999) Autumn migration speed of juvenile reed and sedge warblers in relation to date and fat loads. Condor 101:153–156

Bensch S, Hasselquist D, Hedenström A, Ottosson U (1991) Rapid moult among Palaearctic passerines in West Africa: an adaptation to the oncoming dry season? Ibis 133:47–52

Bibby CJ, Green RE (1981) Autumn migration strategies of reed and sedge warblers. Ornis Scand 12:1–12

Biebach H (1996) Energetics of winter and migratory fattening. In: Carey C (ed) Avian energetics and nutritional ecology. Chapman and Hall, NY

Biebach H (1998) Phenotypic organ flexibility in garden warblers *Sylvia borin* during long-distance migration. J Avian Biol 29:529–535
Brensing D (1989) Ökologische Untersuchungen der Tagesperiodik von Kleinvögeln. Ökol Vögel (Ecol Birds) 11:1–148
Bulyuk VN, Tsvey A (2006) Timing of nocturnal autumn migratory departures in juvenile European robins (*Erithacus rubecula*) and endogenous and external factors. J Ornithol 147: 298–309
Carpenter FL, Hixon MA, Russel RW, Paton DC, Temeles EJ (1993) Interference asymmetries among age-classes of rufous hummingbirds during migratory stopover. Behav Ecol Sociobiol 33:297–304
Cerasale DJ, Guglielmo CJ (2006) Plasma metabolite profiles: effects of dietary phospholipids in a migratory passerine (*Zonotrichia leucophrys gambelii*). Physiol Biochem Zoology 79: 754–762
Chen C-C, Barrow WC Jr, Ouchley K, Hamilton RB (2011) Search behavior of arboreal insectivorous migrants at Gulf coast stopover sites in spring. Wilson J Ornithol 123:347–359
Chernetsov N (1998) Autumn migration strategies of reed warblers (*Acrocephalus scirpaceus*) and sedge warblers (*A. schoenobaenus*) within Europe. In: Costa LT, Costa H, Araújo M, Silva MA (eds) Simpósio sobre Aves Migradoras na Península Ibérica. SPEA e Universidade de Évora, Évora
Chernetsov N (2001) Fuel storage in migrating blackcaps *Sylvia atricapilla* in autumn: which factors are important? In: Both C, Piersma T (eds) The avian calendar: exploring biological hurdles in the annual cycle. Programme and abstracts of 3rd EOU conference, Haren/ Groningen
Chernetsov NS (2003) Ekologiya i povedenie vorobyinykh ptits na migratsionnykh ostanovkakh: postanovka problemy (Stopover ecology and behaviour of passerine migrants: methodological issues of studies). Ornithologia (Moscow) 30:136–146
Chernetsov N (2010) Migratory stopovers of Wrens *Troglodytes troglodytes* on the south-eastern Baltic coast. Avian Ecol Behav 17:13–22
Chernetsov N, Manukyan A (1999a) Feeding strategy of Reed Warblers *Acrocephalus scirpaceus* on migration. Avian Ecol Behav 3:59–68
Chernetsov N, Manukyan A (1999b) Foraging strategies of the Sedge Warbler (*Acrocephalus schoenobaenus*) and the Reed Warbler (*A. scirpaceus*) on migration. Proc Zool Inst 281: 101–106
Chernetsov N, Manukyan A (2000) Foraging strategy of the Sedge Warbler (*Acrocephalus schoenobaenus*) on migration. Vogelwarte 40:189–197
Chernetsov N, Titov N (2000) Design of a trapping station for studying migratory stopovers by capture-mark-recapture analysis. Avian Ecol Behav 5:27–33
Chernetsov N, Titov N (2001) Migratory stopovers of juvenile Blackcaps *Sylvia atricapilla* in autumn: stopover length, fuel deposition rate, and an attempt to predict departure body mass. Avian Ecol Behav 6:27–28
Chernetsov N, Bulyuk VN, Ktitorov P (2007) Migratory stopovers of passerines in an oasis at the crossroad of the African and Indian flyways. Ring Migr 23:243–251
Cuadrado M (1997) Why are migrant robins (*Erithacus rubecula*) territorial in winter?: the importance of the anti-predatory behaviour. Ethol Ecol Evol 9:77–88
Dierschke V (2003) Predation hazard during migratory stopover: are light or heavy birds under risk? J Avian Biol 24:24–29
Dolnik VR (1975) Migratsionnoye sostoyanie ptits (Migratory disposition of birds). Nauka, Moscow
Dolnik VR, Gavrilov VM (1973a) Caloric equivalent of body weight variation in chaffinches *fringilla coelebs*). In: Bykhovskii BE (ed) Bird migrations, ecological and physiological factors. Halstead Press, NY
Dolnik VR, Gavrilov VM (1973b) Energy metabolism during flight of some passerines. In: Bykhovskii BE (ed) Bird migrations, ecological and physiological factors. Halstead Press, NY

Dunn EH (2000) Temporal and spatial patterns in daily mass gain of magnolia warblers during migratory stopover. Auk 117:12–21

Dunn EH (2001) Mass change during migration stopover: a comparison of species groups and sites. J Field Ornithol 73:419–432

Dunn EH (2002) A cross-Canada comparison of mass change in birds during migration stopover. Wilson Bull 114:368–379

Eggers S (2000) Compensatory frugivory in migratory *Sylvia* warblers: geographical responses to season length. J Avian Biol 31:63–74

Ellegren H (1993) Speed of migration and migratory flight lengths of passerine birds ringed during autumn migration in Sweden. Ornis Scand 24:220–228

Fransson T (1995) Timing and speed of migration in North and West European populations of *Sylvia* warblers. J Avian Biol 26:39–48

Fransson T (1998) Patterns of migratory fuelling in whitethroats *Sylvia communis* in relation to departure. J Avian Biol 29:569–573

Fransson T, Weber TP (1997) Migratory fuelling in blackcaps (*Sylvia atricapilla*) under perceived risk of predation. Behav Ecol Sociobiol 41:75–80

Gannes LZ (2001) Comparative fuel use of migrating passerines: effects of fat stores, migration distance, and diet. Auk 118:665–677

Gladwin TW (1963) Increases in weights of *Acrocephali*. Bird Migr 2:319–324

Grandío JM (1998) Comparacion del peso y su incremento, del tiempo de estancia y de la abundancia del carricerin comun (*Acrocephalus schoenobaenus*) entre dos zonas de la marisma de Txingudi (N de España). Ardeola 45:137–142

Grosch K (1995) Die Nahrungszusammensetzung rastender Kleinvögel auf der Halbinsel Mettnau. Diplomarbeit, Universität Bayreuth

Hansson M, Pettersson J (1989) Competition and fat deposition in Goldcrests (*Regulus regulus*) at a migration stopover site. Vogelwarte 35:21–31

Hume ID, Biebach H (1996) Digestive tract function in the long-distance migratory garden warbler, *Sylvia borin*. J Comp Physiol B 166:388–395

Jenni L, Jenni-Eiermann S (1998) Fuel supply and metabolic constraints in migrating birds. J Avian Biol 29:521–528

Jenni L, Schwilch R (2001) Plasma metabolite levels indicate change in body mass in reed warblers *Acrocephalus scirpaceus*. Avian Sci 1:55–65

Jenni-Eiermann S, Jenni L (1991) Metabolic responses to flight and fasting in night-migrating passerines. J Comp Physiol B 161:465–474

Jenni-Eiermann S, Jenni L (1994) Plasma metabolite levels predict individual body-mass changes in a small long-distance migrant, the garden warbler. Auk 111:888–899

Jenni-Eiermann S, Jenni L (2003) Interdependence of flight and stopover in migrating birds: possible effects of metabolic constraints during refuelling on flight metabolism. In: Berthold P, Gwinner E, Sonnenschein E (eds) Avian migration. Springer, Berlin

Johnson AB, Winker K (2008) Autumn stopover near the Gulf of Honduras by Nearctic–Neotropic migrants. Wilson J Ornithol 120:277–285

Karasov WH, Pinshow B (1998) Changes in lean mass and in organs of nutrient assimilation in a long-distance passerine migrant at a springtime stopover site. Physiol Zool 71:435–448

Kelly JF, Delay LS, Finch DM (2002) Density-dependent mass gain by Wilson's warblers during stopover. Auk 119:210–213

Khlebosolov EI (1993) Stereotip kormovogo povedeniya ptits (Stereotype of foraging behaviour in birds). Uspekhi Sovremennoi Biologii 113:717–730

Khlebosolov EI (1996) Obosnovanie modeli odnomernoy ierarkhicheskoy nishi u ptits (Substantiation of the one-dimensional hierarchical niche model in birds). Uspekhi Sovremennoi Biologii 116:447–462

Khlebosolov EI (2005) Kormovoe povedenie kak vidovaya kharakteristika ptits (Foraging behaviour as a species-specific characteristic of birds). Zool Zhurnal 84:54–62

Klaassen M, Biebach H (1994) Energetics of fattening and starvation in the long-distance migratory garden warbler, *Sylvia borin*, during the migratory phase. J Comp Physiol B 164:362–371
Klaassen M, Lindström Å (1996) Departure fuel loads in time-minimizing migrating birds can be explained by the energy costs of being heavy. J Theor Biol 183:29–34
Klaassen M, Lindström Å, Zijlstra R (1997) Composition of fuel stores and digestive limitations to fuel deposition rate in the long-distance migratory thrush nightingale, *Luscinia luscinia*. Physiol Zool 70:125–133
Klein H, Berthold P, Gwinner E (1971) Vergleichende Untersuchung tageszeitlicher Aktivitätsmuster und tageszeitichen Körpergewichtsänderungen gekäfigter und freilebender Grasmücken (*Sylvia*). Oecologia 8:218–222
Kokko H (1999) Competition for early arrival in migratory birds. J Anim Ecol 68:940–950
Ktitorov P, Bairlein F, Dubinin M (2008) The importance of landscape context for songbirds on migration: body mass gain is related to habitat cover. Landscape Ecol 23:169–179
Labocha MK, Hayes JP (2012) Morphometric indices of body condition in birds: a review. J Ornithol 153:1–22
Leist AJ (2007) The importance of fruit to Swainson's thrushes, *Catharus ustulatus*, during fall migration: a field test of plasma metabolite analysis. MSc thesis, Humboldt State University
Liechti F (1995) Modelling optimal heading and airspeed of migrating birds in relation to energy expenditure and wind influence. J Avian Biol 26:330–336
Liechti F (2006) Birds: blowin' by the wind? J Ornithol 147:202–211
Liechti F, Bruderer B (1998) The relevance of wind for optimal migration theory. J Avian Biol 29:561–568
Lindström Å (1991) Maximum fat deposition rates in migrating birds. Ornis Scand 22:12–19
Lindström Å (2003) Fuel deposition rates in migrating birds: causes, constraints and consequences. In: Berthold P, Gwinner E, Sonnenschein E (eds) Avian migration. Springer, Berlin
Lindström Å, Piersma T (1993) Mass changes in migrating birds: the evidence for fat and protein storage re-examined. Ibis 135:70–78
Loria DE, Moore FR (1990) Energy demands of migration on red-eyed vireos, *Vireo olivaceus*. Behav Ecol 1:24–35
Mädlow W (1997) Durchzug und Rastverhalten des Rotkehlchens (*Erithacus rubecula*) im Herbst 1995 auf der Greifswalder Oie: Situation während eines Masseneinzugs. Seevögel 18:75–81
Merom K, Yom-Tov Y, McClery R (2000) Philopatry to stopover site and body condition of transient reed warblers during autumn migration through Israel. Condor 102:441–444
Moore FR (1994) Resumption of feeding under risk of predation: effect of migratory condition. Anim Behav 48:975–977
Moore FR, Aborn DA (1996) Time of departure by summer tanagers (*Piranga rubra*) from a stopover site following spring trans-Gulf migration. Auk 113:949–952
Moore FR, Kerlinger P (1987) Stopover and fat deposition by North American wood-warblers (Parulinae) following spring migration over the Gulf of Mexico. Oecologia 74:47–54
Moore FR, Wang Y (1991) Evidence of food-based competition among passerine migrants during stopover. Behav Ecol Sociobiol 28:85–90
Neto JM, Encarnação V, Fearon P, Gosler AG (2008) Autumn migration of Savi's warblers *Locustella luscinioides* in Portugal: differences in timing, fuel deposition rate and non-stop flight range between the age classes. Bird Study 55:78–85
Ottich I, Dierschke V (2003) Exploration of resources modulates stopover behaviour of passerine migrants. J Ornithol 144:307–316
Panov IN, Chernetsov NS (2010) Migratsionnaya strategiya varakushki (*Luscinia svecica*) v Vostochnoy Fennoskandii. Soobschenie 1: Osnovnye parametry migratsionnykh ostanovok (Migratory strategy of bluethroats, *Luscinia svecica*, in Eastern Fennoscandia. Part 1: Main stopover parameters). Proc Zool Inst 314:93–104

Pettersson J (1983) Rödhakens *Erithacus rubecula* höstflyttning vid Ottenby. Vår Fågelvärld 42:333–342

Piersma T (1990) Pre-migratory 'fattening' usually involved more than the deposition of fat alone. Ring Migr 11:113–115

Piersma T (1998) Phenotypic flexibility during migration: physiological optimization contingent on the risks and rewards of fuelling and flight. J Avian Biol 29:511–520

Piersma T, Gill RE (1998) Guts don't fly: small digestive organs in obese bar-tailed godwits. Auk 115:196–203

Piersma T, Dietz MW, Ekinga AD, Nebel S, Van Gils J, Battley PF, Spaans B (1999) Reversible size-changes in stomachs of shorebirds: when, to what extent, and why? Acta Ornithol 34:175–181

Rappole JH, Warner DW (1976) Relationships between behavior, physiology and weather in avian transients at a migration stopover site. Oecologia 26:193–212

Salewski V (1999) Untersuchungen zur Überwinterungsökologie paläarktischer Singvögel in Westafrika unter besonderer Berücksichtigung der Wechselwirkungen zu residenten Arten. W&T Verlag, Berlin

Salewski V, Schaub M (2007) Stopover duration of Palearctic passerine migrants in the western Sahara—independent of fat stores? Ibis 149:223–236

Salewski V, Almasi B, Heuman A, Thoma M, Schlageter A (2007) Agonistic behaviour of Palaearctic passerine migrants at a stopover site suggests interference competition. Ostrich 78:349–355

Sapir N, Tsurim I, Gal B, Abramsky Z (2004) The effect of water availability on fuel deposition of two staging *Sylvia* warblers. J Avian Biol 35:25–32

Schaub M, Jenni L (2000) Fuel deposition of three passerine bird species along migration route. Oecologia 122:306–317

Schaub M, Jenni L (2001) Variation of fuelling rates among sites, days and individuals in migrating passerine birds. Funct Ecol 15:584–594

Schaub M, Liechti F, Jenni L (2004) Departure of migrating European robins, *Erithacus rubecula*, from a stopover site in relation to wind and rain. Anim Behav 67:229–237

Schmidt-Nielsen K (1997) Animal physiology: adaptation and environment, 5th edn. Cambridge University Press, NY

Schwilch R, Grattarola A, Spina F, Jenni L (2002) Protein loss during long-distance migratory flight in passerine birds: adaptation and constraints. J Exp Biol 205:687–695

Schwilch R, Jenni L (2001) Low initial refueling rate at stopover sites: a methodological effect? Auk 118:698–708

Seewagen CL, Slayton EJ (2008) Mass changes of migratory landbirds during stopovers in a New York City park. Wilson J Ornithol 120:296–303

Stevens L (2004) Avian biochemistry and molecular biology. Cambridge University Press, Cambridge

Szulc-Olech B (1965) The resting period of migrant robins on autumn passage. Bird Study 12:1–7

Tellería JL, Péres-Tris J (2004) Consequences of the settlement of migrant European robins *Erithacus rubecula* in wintering habitats occupied by conspecific residents. Ibis 146:258–268

Titov N (1999a) Individual home ranges of Robins *Erithacus rubecula* at stopovers during autumn migration. Vogelwelt 120:237–242

Titov N (1999b) Fat level and temporal pattern of diurnal movements of Robins (*Erithacus rubecula*) at an autumn stopover site. Avian Ecol Behav 2:89–99

Titov N (2000) Interaction between foraging strategy and autumn migratory strategy in the robin *Erithacus rubecula*. Avian Ecol Behav 5:35–44

Titov N, Chernetsov N (1999a) How body mass should be compared to structural size when calculating condition index? Avian Ecol Behav 3:111–113

Titov NV, Chernetsov NS (1999b) Stokhasticheskie modeli kak novy metod otsenki prodolzhitelnosti migratsionnykh ostanovok ptits (Stochastic models as a new method for estimating length of migratory stopovers in birds). Uspekhi Sovremennoi Biologii 119:396–403

Tsvey AL (2008) Strategii migratsii zaryanki (*Erithacus rubecula*) v vostochnoy Pribaltike (Migratory strategies of the European robin (*Erithacus rubecula*) in the eastern Baltic). Dissertation, Zoological Institute RAS

Tsvey A, Bulyuk VN, Kosarev V (2007) Influence of energy condition and weather on departures of first-year European robins, *Erithacus rubecula*, from an autumn migratory stopover site. Behav Ecol Sociobiol 61:1665–1674

von Stünzner-Karbe D (1996) Territorialität, Habitatnutzung und Furagierverhalten überwinternder Trauerschnäpper (*Ficedula hypoleuca*) in West-Afrika. Diplomarbeit, Universität Bayreuth

Wang Y, Moore FR (1997) Spring stopover on intercontinental migratory thrushes along the northern coast of the Gulf of Mexico. Auk 114:263–278

Wang Y, Moore FR (2005) Long-distance bird migrants adjust their foraging behavior in relation to energy stores. Acta Zoologica Sinica 51:12–23

Williams TD, Guglielmo CG, Egeler O, Martyniuk CJ (1999) Plasma lipid metabolites provide information on mass change over several days in captive Western sandpipers. J Avian Biol 33:1–9

Winker K (1995) Autumn stopover on the Isthmus of Tehuantepec by woodland Nearctic–Neotropic migrants. Auk 112:690–700

Winker K, Warner DW, Weisbrod AR (1992) Daily mass gains among woodland migrants at an inland stopover site. Auk 109:853–862

Yablonkevich ML, Shapoval AP (1987) Massa tela, energeticheskie rezervy i povedenie nekotorykh vidov nochnykh migrantov vo vremya dnevnoy ostanovki v period osennego proleta na ostrove Barsakelmes Aralskogo morya (Body mass, energy reserves and behaviour of some nocturnal migrants during at daytime stopover during autumn passage across the Aral Sea island, Barsakelmes). Proc Zool Inst 173:13–47

Yablonkevich ML, Bolshakov CV, Bulyuk VN, Eliseev DO, Efremov VD, Shamuradov AK (1985) Massa tela i zhirnost' ptits, proletayuschikh vesnoy cherez pustyni Sredney Azii (Body weight and fat of birds passing across the Middle Asia deserts in spring). Proc Zool Inst 137:11–59

Yosef R, Wineman A (2010) Differential stopover of blackcap (*Sylvia atricapilla*) by sex and age at Eilat, Israel. Naturwissenschaften 74:360–367

Zimin VB (2003) Body mass variability in juvenile Robins *Erithacus rubecula* in the Ladoga area. Avian Ecol Behav 10:1–31

Chapter 4
Optimal Migration Theory

Abstract This chapter is devoted to the critical analysis of the optimal migration theory. I suggest that this theory has multiple issues, mainly because some of its basic assumptions are not supported by the empirical evidence. The fundamental assumption of the U-shaped flight power curve is not supported by the data. In the birds with high aerodynamic quality (most passerine migrants belong to this category) the flight costs are independent of flight speed, within a rather broad range of flight speed routinely employed. Nevertheless, the optimal migration theory played a very important role in the attempts to find quantitative relationships between the main ecological parameters of migratory stopovers, i.e. between stopover duration, fuel deposition rate and departure fuel load. It has advanced stopover studies greatly, but I claim that its critical revision is necessary.

4.1 General Remarks

The optimal migration theory aims to establish quantitative relationships between the main energetic parameters of migration, i.e. stopover duration, fuel deposition rate and departure fuel load. It should be emphasized that in spite of its very general name, which apparently lays claim to summarising all parameters of avian migration, concerning both flights and stopovers, this theory only deals with quantitative energetic parameters and completely ignores such important aspects of migratory behaviour as habitat selection and use, spatial behaviour at stopovers, and diel pattern of migratory flights. Obviously, a characteristic of flight and stopover behaviour of avian migrants which tries to be comprehensive cannot omit these factors; therefore the optimal migration theory which simply has no room for these aspects is incomplete.

This theory started from the famous and much cited paper by Alerstam and Lindström (1990). The authors used the methodology of behavioural ecology, the

N. Chernetsov, *Passerine Migration*, DOI: 10.1007/978-3-642-29020-6_4,

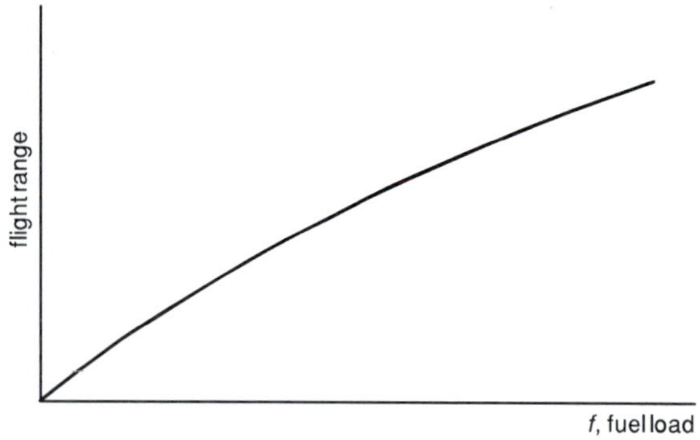

Fig. 4.1 The potential flight distance as a function of fuel load (*f*), from Eq. 4.1

optimal foraging theory to be more exact (Begon et al. 2006), to analyse behavioural strategies of migrating birds at stopover. It has been assumed that during migration, birds should optimise their behaviour to minimise the expenditure of one of the three currencies: the time spent on travelling, total energy expenditure for migration, and predation risk. Apparently the most realistic assumption would be minimisation of the mortality rate during migration. However, there are no realistic quantitative estimates of mortality risk for individual migrants, and it is not even clear how to obtain such estimates before satellite tracking of small songbirds becomes commonplace (Robinson et al. 2010). It is difficult to imagine how to relate the survival probability to the known energetic factors. Therefore, the authors suggested the following three currencies (Alerstam and Lindström 1990).

They considered variants of behaviour that should lead to time and energy minimisation. The authors based their calculations on the basic relationship that predicts the possible flight range from the fuel load (Fig. 4.1):

$$Y(f) = c\left(1 - \frac{1}{(1+f)^{0.5}}\right) \tag{4.1}$$

where c is the constant with the dimension of length which depends on the aerodynamic constant of the bird (i.e. its aerodynamic quality), fuel composition (fat to protein ratio) and lean body mass, and f is the relative fuel load [$f = (m - m_0)/m_0$ where m is the bird's body mass and m_0 is its fat-free body mass (Alerstam and Lindström 1990; Alerstam and Hedenström 1998)]. The Eq. 4.1 forms the basis of the optimal migration theory by predicting that flight range is a diminishing return function of fuel load. This equation is the cornerstone on which the relationships predicting optimisation rules, depending on choosing the main currency or their combination, are based. Below we shall discuss the importance of the Eq. 4.1 for the validity of the optimal migration theory.

It is worth noting that the authors of the optimal migration theory consider in great detail the behaviour of migrants that are minimising migration time (i.e. maximising migration speed), in slightly less detail the behaviour of energy minimisers, and just mention the possibility rather than analyse the strategy of predation risk minimisation.

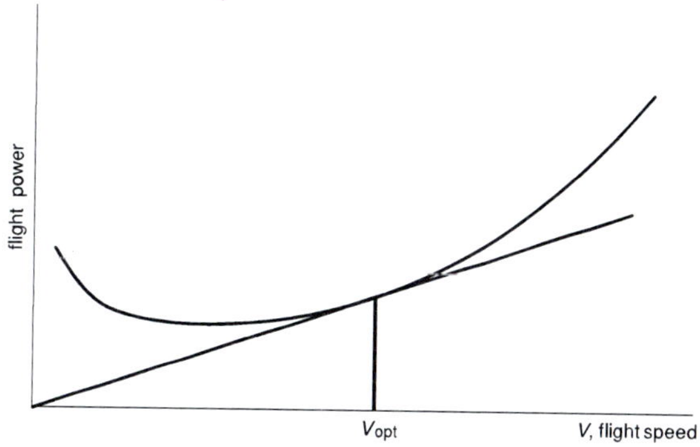

Fig. 4.2 The flight power as a function of flight speed predicted from the aerodynamic theory. V_{opt} is the optimal migration speed determined from the tangent to the flight power curve from the point of origin

4.2 Time Minimisation

The relationship between flight range and fuel load is positive, but it is a diminishing return function (Eq. 4.1). It is assumed on the basis of the idea that transport of extra fuel has energy cost, i.e. the heavier a bird is, the higher is its flight cost. In its turn, this notion is based on the assumption based on aerodynamic flight theory that flight power curve (the relationship between flight cost and flight speed) is U-shaped (Fig. 4.2; Pennycuick 1975, 1989; Dolnik 1995; Engel et al. 2010). The higher is the fuel load of a bird the less additional fuel adds to its potential flight range.

Assuming that migratory behaviour is optimised in the manner to cover the migratory route as quickly as possible, the birds should terminate their stopover (i.e. to start the next migratory flight) when their fuel load reaches the threshold at which the instantaneous migratory speed (the speed that can be reached with the given fuel load) drops to the mean optimal speed of migration *V*(*t*). This rule is analogous to the marginal value theorem in the optimal foraging theory (Charnov 1976).

The mean speed of migration is found from the following relationship:

$$V(t) = \frac{Y(t)}{t + t_0} = \frac{c}{t + t_0}\left(1 - \frac{1}{(1 + kt)^{0.5}}\right) \tag{4.2}$$

with $f = k \times t$, where k is fuel deposition rate, t is duration of refuelling period, and t_0 is search/settling time (i.e. the stopover duration is $t_0 + t$). The relationship between migratory speed and stopover duration under given FDR is shown in Fig. 4.3. Obviously, the minimisation of time spent on migration is equivalent to maximising migration speed. Our task of calculating optimal stopover time t^* and thus optimal fuel load f^* is thus reduced to finding the value of t under which the function *V*(*t*) reaches its maximum. It is necessary to find such t values under which the derivative dV/dt is either equal to zero or does not exist. In other words, it is necessary to solve the equation

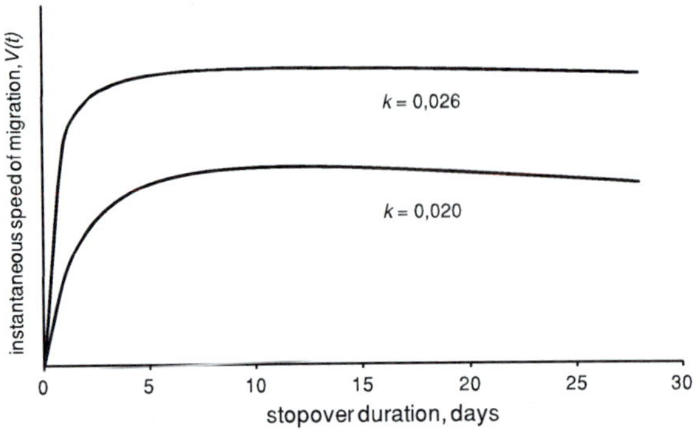

Fig. 4.3 The speed of migration $V(t)$ as a function of stopover duration t for two different fuel deposition rate k, $c = 6{,}000$, $t_0 = 1$

$$\frac{dY}{dt} = \frac{Y(t)}{t + t_0} \tag{4.3}$$

or, which is equivalent,

$$\frac{1}{2}k(1 + kt)^{-3/2} = \frac{1}{t + t_0}\left(1 - (1 + kt)^{-0.5}\right) \tag{4.4}$$

It may be formally written as

$$t^* = \underset{t}{\arg} \max_{t > 0} V(t) \tag{4.5}$$

Assuming that fuel deposition rate k is constant throughout the stopover (after the end of the search/settling time t_0 during which energy f_0 is used; see Sect. 3.5) and keeping in mind that $f = k \times t$, the optimal departure fuel load (i.e. fuel load upon reaching which it is optimal to take off) may be calculated as $f^* = k \times t^*$ (Alerstam and Lindström 1990; Lindström and Alerstam 1992; Hedenström and Alerstam 1997). The time of the migratory flights is neglected, first, because it is small compared to stopover time, and second, because in nocturnal migrants flights are performed at night, i.e. they take time from sleeping but not from foraging.

The optimal departure fuel load as a function of FDR is shown in Fig. 4.4. Assuming the search/settling time (t_0) of 2 days, which is a realistic assumption (see Sect. 3.5), the optimal stopover time would be 10.6 days when $k = 0.04$ and 7.7 days when $k = 0.08$. The optimal departure fuel load (f^*) would then be 0.42 and 0.62, respectively. It appears that time minimising migrants (i.e. those that maximise their migration speed) should stop over for shorter time at optimal stopover sites than at suboptimal sites that permit slower fuel deposition rates. In spite of shorter stopovers they gain more fuel at optimal sites. This relationship is continuous if migration distance is infinitely large. If a more realistic assumption of a finite migration distance is made, the optimal migration distance will grow with increasing FDR stepwise (Weber and Houston 1997b). Therefore, the theory predicts that at the best sites (with the highest FDR) migrants should gain more fuel than is necessary to reach the next potential stopover site. Less optimal sites, conversely, should be skipped in order not to waste time at them (Weber et al. 1994).

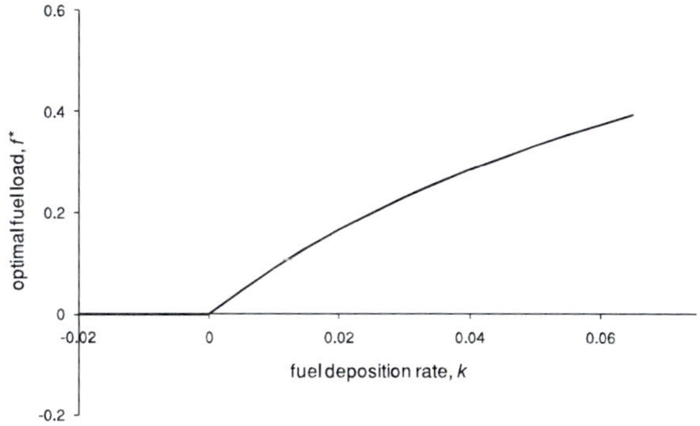

Fig. 4.4 The optimal fuel load f^* as a function of fuel deposition rate k

Several attempts have been made to test whether the behaviour of migrants (to be more exact, the form of relationship between energetic stopover parameters) conforms to the predictions of the optimal migration theory for time minimisers. These tests were mainly made by experimental manipulation of FDR by providing additional food for migrants at stopovers (Lindström and Alerstam 1992; Fransson 1998; Dänhardt and Lindström 2001; Bayly 2006). The field data agreed with theoretical predictions qualitatively but not quantitatively (Lindström and Alerstam 1992; Fransson 1998; Bayly 2006), and in one study no significant relationship between FDR and departure fuel load was found at all (Dänhardt and Lindström 2001).

It should be however noted that in this latter experiment departure fuel loads of experimental European robins (on average 0.53) were much higher than those recorded in their conspecifics on the Courish Spit without supplementary feeding (Chernetsov et al. 2004; Bulyuk and Tsvey 2006). Our data on migrating European robins that received no supplementary food indicate a positive relationship between FDR and departure fuel load (Chernetsov et al. 2004). A detailed analysis of the relationship between these parameters in robins that stopped over for three and more days showed a significant positive correlation (Bulyuk and Tsvey 2006). However, the authors correctly mention that only 51% of individuals included into their analysis fell into this category. It remains obscure which (optimisation?) criteria were used by the remaining 49% of European robins when making their departure decisions.

To explain the discrepancies between the empirical results and the predictions of the optimal migration theory, it has been suggested that (1) the expected mean migratory speed varies between individuals (Lindström and Alerstam 1992); (2) the stepwise relationship between departure fuel load and FDR suggested by Weber and Houston (1997b) means no relationship between these parameters within the steps; (3) when high fuel loads are reached, effective FDR decreases because of the significant metabolic costs of living with high fat deposits (Klaassen and Lindström 1996); (4) migrants minimise not migration time but its energy cost (Hedenström and Alerstam 1997). Further development of the optimal migration theory included the pattern of variation of FDR along the migratory route. The local variation is the situation in which the birds travel from the stopover site with FDR of k_0 to the next site with the FDR of k_1 and expect that during the future movements they will encounter the initial value of k_0, and k_1 was an exception. The global variation is the situation in which the migrants encounter k_1 and expect to have this situation further along their route, i.e. when they update their expectations.

This line of reasoning looks a very scholastic one because it is difficult to see how to test (i.e. to falsify) the hypotheses speculating on what the animals expect. However, this is an important argument because the assumption of the local variation results in the prediction of a much steeper relationship of f_{dep} on k than the assumption of the global variation (Fig. 2 in Houston 1998). In practice it means that practically any empirical result may be explained in the framework of the optimal migration theory, if respective assumptions are made *a posteriori*. It is important to note that one assumption will be apparently not worse than the other one.

Chernetsov et al. (2004) have suggested that the most realistic assumption is not local or global, but the stochastic variation of FDR along the migratory route, when the empirical FDR value is compared with the migrant with the expected mean and its variance. It may be compared with the behaviour of a human job seeker who compares job offers (and salaries!) with what they expect on the basis of the market value of their qualifications. The mean FDR expected by the migrants may be based on their previous experience ('familiarity with the market situation' in our metaphor), as we suggested (Chernetsov et al. 2004) or it may be endogenous, as assumed by Bayly (2006). Neither it can be ruled out that stopover duration varies around the fixed mean value which was shaped by the selection on the basis of usually occurring fuel deposition rates (Erni et al. 2002; Bayly 2006). It is easy to see that the latter assumption (which looks rather realistic and fits the existing empirical data not too badly) is beyond the framework of the optimal migration theory.

It should be also kept in mind that one of the basic assumptions of the optimal migration theory is that after the search/settling period which may be characterised by low or even negative FDR (see Sect. 3.5), the rate of gaining mass remains constant (Hedenström and Alerstam 1997; Weber and Houston 1997b). This assumption is generally not held, as shown by the data on FDR of sedge warblers at the feeding table (Bayly 2007). As FDR declines towards the end of stopover, the prediction of positive relationship between FDR and departure fuel load in time minimisers is also not fulfilled.

4.3 Minimisation of Energy Cost of Migration

In the previous section, I have given some arguments that, in my opinion, speak against the validity of time minimisation assumption. It should be emphasized that optimal migration theory does not *predict* time or energy minimisation. It *assumes* them *a priori* and makes specific predictions on the basis of these assumptions.

The most general argument against the assumption of time minimisation is that it is not very realistic. It is usually believed that in autumn, unlike spring, the speed of migration cannot be of utmost importance, because migrants are not faced with the necessity to occupy the breeding territory first (Hedenström and Alerstam 1997). For some migrants it is not correct, e.g. for pied flycatchers that occupy wintering territories, and the first arriving individuals have access to top quality plots (von Stünzner-Karbe 1996; Salewski 1999; Salewski and Jones 2006). Neither is the spring situation one-dimensional. There is a premium on early arrival, but there may be very significant costs of being too early in the years with cold spells. Spring mortality in some years and species of passerines may be very significant (Payevsky 1985). Therefore, it is not self-evident that optimal migratory speed is the maximum possible one, even during spring migration.

In this context the assumption that migrating passerines optimise their behaviour to minimise the energy spent on migration in many cases looks more realistic. One should distinguish between the cost of transport of fuel and the total energy cost of migration (Hedenström and Alerstam 1997). If the aim is to migrate spending the minimum energy possible per unit distance, it is necessary to find the local maximum for the following function:

$$R = \frac{Y(f) - Y(f_0)}{f} \tag{4.6}$$

where f_0 is the energy cost of search/settling. By differentiating Eq. 4.6 by f and equating the derivative to zero we obtain

$$\frac{dY}{df} = \frac{Y(f) - Y(f_0)}{f} \tag{4.7}$$

which allows us to find f^*. It is noteworthy that f^* depends only on f_0, but is independent of k or t_0.

Independence of optimal departure fuel load on FDR is the most characteristic feature of the model based on the assumption of minimising energy spent per unit distance (Alerstam and Lindström 1990). Obviously, to reach the same departure fuel load a migrating bird would need more time at a site with low k than at a site with high k, but whatever low FDR at a given site would be (given that it is positive), a long stopover at such site is predicted to be optimal. This situation could only be realistic if time allocated for migration is infinite. In the real world where infinitely low speed of migration is apparently unacceptable, the assumptions of the model must be modified.

4.4 Predation Risk Minimisation

As already mentioned, safety as the main optimisation currency was mentioned and briefly discussed in the first seminal publication on optimal migration theory (Alerstam and Lindström 1990), but hardly elaborated in the subsequent development. The reason probably is that this parameter is very difficult to quantify, and to obtain its more or less reliable estimates is even less easy. It might be assumed that fatter birds would be in greater danger of predation than their leaner conspecifics because of the poorer manoeuvrability of the former. This assumption was made by the authors of the optimal migration theory (Alerstam and Lindström 1990) who concluded that in such a case the optimal departure fuel load f^* should be lower than under time minimisation assumption. However, a field study of this topic (Dierschke 2003) showed that among songbird migrants on Helgoland in the North Sea among victims of feral cats and birds of prey (mainly sparrowhawks *Accipiter nisus*) lean and not fat migrants were overrepresented, as assumed by Alerstam and Lindström (1990).

It means that fuel loads that routinely occur among migrating passerines not facing major ecological barriers do not challenge their manoeuvrability and the ability to escape predators. However, lean individuals that need to refuel urgently, are often forced to utilise risky foraging tactics (e.g. to forage in open habitats with little or no cover, or to compromise their vigilance) and because of that are taken by predators disproportionally often. Direct visual observations of foraging behaviour of red-eyed vireos in spring after migratory flights across the Gulf of Mexico showed that lean birds used a wider array of foraging manoeuvres and substrates and probably were under greater risk of predation than their fatter conspecifics (Loria and Moore 1990).

Even among passerines that prepare to cross a serious barrier and deposit large fuel stores, the risk of predation only slightly increases with fuel load. Theoretical calculations show that if a songbird with a fuel load of 80% of its lean body mass (which is close to the maximum values known for passerines) feeds 0.5 m from the cover, it reaches it by 8% later than a lean bird. At the same time, if this bird discovers a predator by one standard deviation later than the mean, it reaches cover by 53% later (Lind 2004). It means that even in very fat birds the decline in their physical ability to escape is negligible compared to the natural variation in the ability to detect a predator (Lind and Cresswell 2006). The problem may only become acute in birds that forage in very open habitats without any cover, like snow buntings *Plectrophenax nivalis* or waders—but even in such cases flocking behaviour may provide a solution. Another matter is that the presence of predators at a stopover site influences the behaviour of migrants: they may select more covered microhabitats and possibly decrease their FDR (Moore 1994; Cimprich et al. 2005).

4.5 Basic Equations

Apart from the aforementioned difficulties that mainly refer to the biological validity of the assumptions of the optimal migration theory, there is also the issue of the validity of the basic relationships, i.e. of the flight range equation (Eq. 4.1) and flight power curve (Fig. 4.2). The flight range equation is one of the two main assumptions on which the optimal migration theory is based (Alerstam and Lindström 1990; Hedenström 2008). Weber and Houston (1997a) showed that using different flight power estimates leads to the following generalisation of this equation:

$$Y(f) = c[1 - \left(1 + f\right)^{-\zeta}], \quad \text{with } 0 < \zeta < 1 \tag{4.8}$$

In any case, the flight range equation remains a diminishing return function, i.e. the more fuel a migrant deposits the less is the flight range added with the same absolute amount of the fuel added. The reason is the high cost of transport of additional fuel.

Another relationship of profound importance for optimal migration theory is the one between flight speed and the power of flight. This relationship is widely accepted as being described by the following formula:

$$P = \alpha + \beta \times V^{-1} + \gamma \times V^3 \tag{4.9}$$

where P is flight power, v is flight speed, and α, β, and γ are physical parameters of the air and the bird (Pennycuick 1989; Hedenström 2002, 2008; Engel et al. 2010). This function is U-shaped (Pennycuick 1975, 1989; Fig. 4.2), which means that there is a single minimum power speed value, above and below which energy expenditure increases.

Both of these crucial relationships (Eqs. 4.1 or 4.8 and 4.9) are based on mechanical flight theory (Pennycuick 1975, 1989; Lindhe Norberg 2004; Hedenström 2008). However, neither of them is fully supported by recent experimental data.

Four recent studies measured energy costs of several hours of flight in wind tunnels in relation to intraspecific variation in body mass (Kvist et al. 2001; Engel et al. 2006; Schmidt-Wellenburg et al. 2007, 2008). Flight costs in rosy starlings *Sturnus roseus* were shown to increase with $m^{0.55}$ [95% confidence interval (CI) of the scaling exponent: 0.36–0.75; Engel et al. (2006)]. In another study of the same species, the scaling exponent was 0.57 (95% CI: 0.40–0.74) or 0.47 (95% CI: 0.18–0.76), depending on the experimental conditions (Schmidt-Wellenburg et al. 2008). In the barn swallow *Hirundo rustica*, the scaling exponent was 0.58 (95% CI: 0.19–0.97; Schmidt-Wellenburg et al. 2007), and in the red knot *Calidris canutus* it was 0.35 (95% CI: 0.08–0.67; Kvist et al. 2001).

All these values are clearly below 1; that is, flight costs increase much less steeply than predicted by the current aerodynamic theory (scaling exponent of 1.1–1.6; Pennycuick 1975; Norberg 1990, 1996; Rayner 1990) and than predicted from among-species allometric equations (0.7–1.9; Norberg 1996; Butler and Bishop 2000; Rayner 1990; Videler 2005). Apparently, only within-species comparisons are relevant for the optimal migration theory, whereas among-species comparisons are not.

The flight range equation can be obtained by integrating the flight power equation $\frac{dY}{dM} = \frac{V(M) \times E}{P(M)}$ [Eq. 9 from Weber and Houston (1997a)], where $V(M)$ is mass-dependent flight speed, E is energy density of fuel stores, a $P(M)$ is the flight range equation. If we ignore the mass dependence of flight speed, this means that

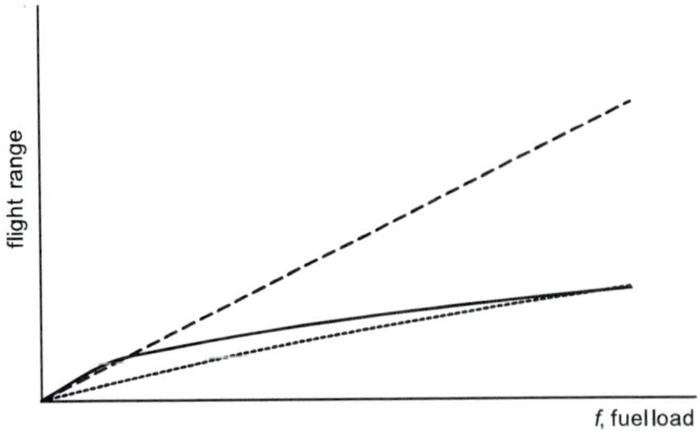

Fig. 4.5 The potential flight range as a function of fuel load. The *dotted line* is the flight range proportional to $f^{-0.5}$ (from Eq. 3.1, Fig. 4.1); the *solid line* represents the $f^{0.5}$ relationship; the *dashed line* shows the direct linear relationship from f

$Y(f)$ is proportional to $(1 + f)^{\zeta}$, where $\zeta > 0$ (cf. Eq. 4.8). Similarity, Delingat et al. (2008) assumed on the basis of empirical measurements that body mass loss in migratory flight is ca. 1% of body mass per hour. This is equivalent to $Y(f)$ proportional to ln $(1 + f)$. Clearly, the calculations of flight and stopover parameters are affected by these results, because both these relationships deviate from the linear function much less than Eq. 4.1 (Fig. 4.5).

Most reviews of avian flight physiology have claimed that the U-shaped relationship between flight speed and flight power is well supported by the experimental data (Norberg 1996; Schmidt-Nielsen 1997, Blem 2000, Harrison and Roberts 2000). Only Ellington (1991) emphasized the scarcity of data to support this claim. In fact, until recently, the U-shaped relationship predicted by mechanical flight theory has been supported by a single study involving a single species, the budgerigar *Melopsittacus undulates* (Tucker 1968), whereas other studies showed either a weak positive relationship, as in the laughing gull *Leucophaeus atricilla* (Tucker 1972), fish crow *Corvus ossifragus* (Bernstein et al. 1973), and European starling *Sturnus vulgaris* (Ward et al. 2004), or similar flight power across a wide range of flight speeds, as in European starling (Torre-Bueno and Larochelle 1978) and geese (Ward et al. 2002). A recent wind-tunnel study showed independence of flight costs from flight speed in rosy starlings despite a 55% increase in flight speed (Engel et al. 2006). A clearly U-shaped relationship between flight power and speed was shown by in vivo measurements of muscle force in cockatiels *Nymphicus hollandicus*, whereas the relationship was weakly U-shaped in ringed turtle-doves *Streptopelia risoria* (Tobalske et al. 2003) and flat in black-billed magpies *Pica pica* (Dial et al. 1997). Recently, a U-shaped relationship was reported for the budgerigar and cockatiel (Bundle et al. 2007). It is noteworthy that both cases of the unequivocally U-shaped relationship involved species that do not migrate to long distances and generally do not fly a lot.

Berger (1985) found J-shaped relationships in the sparkling violetear *Colibri coruscans* and green violetear *C. thalassinus*: their flight metabolism did not vary significantly between hovering speeds and up to 7 m s^{-1} (which is a quite significant speed for a bird of this size). At even higher flight speeds, the metabolic rate increased.

In the recent years, new data on migratory speed of wood thrushes *Hylocichla mustelina*, purple martins *Progne subis* (Stutchbury et al. 2009, 2011) and Arctic turns (Egevang et al. 2010) have been obtained by geolocators. The migratory speed of these three species appeared to be considerably higher than predicted by the optimal migration theory for the birds of their size. Even though an attempt was made to explain this discrepancy between theoretical predictions and empirical measurements by higher than assumed FDR (Bowlin et al. 2010), in the light of the aforementioned data it seems more plausible that the main reason is lower than the hitherto assumed cost of transport of additional fuel, i.e. a larger flight range with given fuel stores.

We are forced to conclude that both main theoretical assumptions on which the optimal migration theory is based are not supported by most empirical data [see also Chernetsov 2010 but see objections by Hedenström (2012)]. When fuel stores are low, the relationship between potential flight range and relative fuel stores deviates from the linear proportionality much less than assumed by Eq. 4.1 (Fig. 4.5). Under such conditions, the cost of transport of additional fuel is low, but it becomes higher and in better agreement with the predictions of the current aerodynamic theory with increasing fuel stores. This is supported by the data on escape flights of blackcaps (Kullberg et al. 1996) and sedge warblers (Kullberg et al. 2000). Kullberg et al. (1996: Fig. 3, 2000: Fig. 1) claimed that flight speed and acceleration decreased with increasing fuel load, but this effect was apparent only when the fuel load exceeded 30% of lean body mass. This means that when $f < 0.25$–0.30, transport of additional fuel is nearly free, i.e. the flight range is nearly directly proportional to fuel stores. Even though these studies are usually cited as supporting the claim that flight maneuverability of fat songbirds is decreased (Kullberg et al. 1996, 2000; Lind et al. 1999), a detailed inspection of graphs clearly shows that it is only true when the fuel load exceeds 30%. In European robins the fattest bird carried only 27% of fuel, and escape flight speed was not related to the fuel load (Lind et al. 1999).

There is no doubt that birds with large fuel stores, exceeding 25–30% of lean body mass, spend more energy for flight than lean individuals. The data available, however, strongly suggests that this effect is limited to individuals with considerable fuel stores (e.g. in sedge warblers, Kullberg et al. 2000). The current calculations of optimal migration theory may approach reality for birds that cross large ecological barriers (e.g. the Sahara or the Gulf of Mexico) and carry large fat stores, but this is an interesting special case of avian long-distance migration. Most passerines that migrate over suitable habitats with continuous stopover possibility usually carry moderate fuel stores <30% of their lean body mass (Bairlein 1987, 1991; Table 4.1), which means that most songbird migrants not crossing an ecological barrier remain in the zone of nearly free transport of extra fuel, where their potential flight range is nearly directly proportional to their fuel stores.

One cannot but agree with Victor Dolnik who wrote that 'aerodynamic models that assume just several constant (aerodynamic) parameters for a flying bird and apply fixed-wing theory to avian flight that occurs in a non-stationary regime with changing geometry and kinematics are insufficient for predicting flight power'

Table 4.1 The proportion of birds with fuel load > 25% of their lean body mass ($f > 0.25$) in captures on the Courish Spit during the spring and autumn passage (1993–2006)

Species	Spring	Autumn
European robin	0.69% (n = 29,285)	0.81% (n = 44,501)
Eurasian reed warbler	0.69% (n = 3,938)	19.4% (n = 2,750)
Sedge warbler	5.9% (n = 2,884)	6.2% (n = 3,327)
Blackcap	1.5% (n = 2,890)	14.3% (n = 5,967)
Garden warbler	0.57% (n = 1,056)	6.9% (n = 2,037)
Willow warbler	0.38% (n = 2,121)	1.8% (n = 5,502)

(Dolnik 1995, p. 94). An additional mass, which is either evenly distributed over the body or with an emphasis on the center of gravity, may even enhance the lift of the body and tail and only slightly impair drag (Dolnik 1995).

4.6 Concluding Remarks

The aforementioned considerations seriously undermine the optimal migration theory in its current form. However, it would not be right to say that this avenue of research has been a dead end. The idea of finding quantitative relationships that govern stopover behaviour of migrants (for how long to stop over; when to take off; how much fuel to deposit) is a most enticing one. Our idea of relationships between energy parameters of stopovers has been greatly improved. The problem with the optimal migration theory is that it attempts to find quantitative relationships on the basis of very incomplete parameters. In reality, departure decisions are made by the migrants on the basis of multiple factors (see Chap. 8). At least as important as energetic considerations (that are studied by the optimal migration theory) are the weather (Åkesson and Hedenström 2000) and possibly the calendar factor, i.e. the position of a migrant on the migratory route with respect to the optimal timing of migration. Omitting these factors from the models of stopover behaviour makes these models incomplete and unrealistic. On the other hand, when these parameters (many of which are very difficult to quantify and even more difficult to estimate reliably) are included, the model gets so many degrees of freedom that it makes little sense to compare its predictions with the empirical reality. In other words, the models become exceedingly difficult to falsify. This is why the attempts to modify the optimal migration theory by including most of the aforementioned parameters and parallel use of increasingly sophisticated modelling algorithms (like e.g. Pareto analysis; Vrugt et al. 2007) can be hardly called successful. Very sophisticated modelling methods can only be used with very simplistic basic assumptions (e.g. assuming equal FDR in all birds at all sites within 0.5° × 0.5° grid cells etc.; Vrugt et al. 2007), which takes us back to the problem of non-realistic assumptions.

References

Åkesson S, Hedenström A (2000) Wind selectivity of migratory flight departures in birds. Behav Ecol Sociobiol 47:140–144

Alerstam T, Hedenström A (1998) The development of bird migration theory. J Avian Biol 29:343–369

Alerstam T, Lindström Å (1990) Optimal bird migration: the relative importance of time, energy, and safety. In: Gwinner E (ed) Bird migration. Springer, Berlin

Bairlein F (1987) The migratory strategy of the garden warbler: a survey of field and laboratory data. Ring Migr 8:59–72

Bairlein F (1991) Body mass of Garden Warblers (*Sylvia borin*) on migration: a review of field data. Vogelwarte 36:48–61

Bayly NJ (2006) Optimality in avian migratory fuelling behaviour: a study of a trans-Saharan migrant. Anim Behav 71:173–182

Bayly NJ (2007) Extreme fattening by sedge warblers, *Acrocephalus schoenobaenus*, is not triggered by food availability alone. Anim Behav 74:471–479

Begon M, Townsend CR, Harper JL (2006) Ecology: from individuals to ecosystems, 4th edn. Blackwell, Oxford

Berger M (1985) Sauerstoffverbrauch von Kolibris (*Colibri coruscans* und *C. thalassinus*) beim Horizontalflug. In: Nachtigall W (ed) BIONA report 3. G Fischer, Stuttgart

Bernstein MH, Thomas SP, Schmidt-Nielsen K (1973) Power input during flight in the fish crow *Corvus ossifragus*. J Exp Biol 58:401–410

Blem CR (2000) Energy balance. In: Whittow GC (ed) Sturkie's avian physiology. Academic Press, San Diego

Bowlin MS, Bisson I-A, Shamoun-Baranes J, Reichard JD, Sapir N, Marra PP, Kunz TH, Wilcove DS, Hedenström A, Guglielmo CG, Åkesson S, Ramenofsky M, Wikelski M (2010) Grand challenges in migration biology. Integr Comp Biol 50:261–279

Bulyuk VN, Tsvey A (2006) Timing of nocturnal autumn migratory departures in juvenile European robins (*Erithacus rubecula*) and endogenous and external factors. J Ornithol 147:298–309

Bundle MW, Hansen KS, Dial KP (2007) Does the metabolic rate-flight speed relationship vary among geometrically similar birds of different mass? J Exp Biol 210:1075–1083

Butler PJ, Bishop CM (2000) Flight. In: Whittow GC (ed) Sturkie's avian physiology. Academic Press, San Diego

Charnov EL (1976) Optimal foraging, the marginal value theorem. Theor Popul Biol 9:129–136

Chernetsov N (2010) Recent experimental data on the energy costs of avian flight call for a revision of optimal migration theory. Auk 127:232–234

Chernetsov NS, Skutina EA, Bulyuk VN, Tsvey AL (2004) Optimal stopover decisions of migrating birds under variable stopover quality: model predictions and the field data. Zh Obschei Biol 65:211–217

Cimprich DA, Woodrey MS, Moore FR (2005) Passerine migrants respond to variation in predation risk during stopover. Anim Behav 69:1173–1179

Dänhardt J, Lindström Å (2001) Optimal departure decisions of songbirds from an experimental stopover site and the significance of weather. Anim Behav 62:235–243

Delingat J, Bairlein F, Hedenström A (2008) Obligatory barrier crossing and adaptive fuel management in migratory birds: the case of the Atlantic crossing in northern wheatears (*Oenanthe oenanthe*). Behav Ecol Sociobiol 62:1069–1078

Dial KP, Biewener AA, Tobalske BW, Warrick DR (1997) Mechanical power output of bird flight. Nature 390:67–70

Dierschke V (2003) Predation hazard during migratory stopover: are light or heavy birds under risk? J Avian Biol 24:24–29

Dolnik VR (1995) Resursy energii i vremeni u ptits v prirode (Energy and time resources in free-living birds). Nauka, St Petersburg

Egevang C, Stenhouse IJ, Phillips RA, Petersen A, Fox JW, Silk JRD (2010) Tracking of Arctic terns *Sterna paradisaea* reveals longest animal migration. Proc Natl Acad Sci USA 107:2078–2208
Ellington CP (1991) Limitations on animal flight performance. J Exp Biol 160:71–91
Engel S, Biebach H, Visser GH (2006) Metabolic cost of avian flight in relation to flight velocity: a study in rose-coloured starlings (*Sturnus roseus* Linnaeus). J Comp Physiol B 176:415–427
Engel S, Bowlin MS, Hedenström A (2010) The role of wind-tunnel studies in integrative research on migration biology. Integr Comp Biol 50:323–335
Erni B, Liechti F, Bruderer B (2002) Stopover strategies in passerine bird migration: a simulation study. J Theor Biol 219:479–493
Fransson T (1998) Patterns of migratory fuelling in whitethroats *Sylvia communis* in relation to departure. J Avian Biol 29:569–573
Harrison JF, Roberts SP (2000) Flight respiration and energetics. Annu Rev Physiol 62:179–205
Hedenström A (2002) Aerodynamics, evolution and ecology of bird flight. Trends Ecol Evol 17:415–422
Hedenström A (2008) Adaptations to migration in birds: behavioural strategies, morphology and scaling effects. Phil Trans R Soc B 363:287–299
Hedenström A (2012) Recent experimental data on the energy costs of avian flight do not call for a revision of optimal migration theory. Auk 129 (in press)
Hedenström A, Alerstam T (1997) Optimum fuel loads in migratory birds: distinguishing between time and energy minimization. J Theor Biol 189:227–234
Houston AI (1998) Models of optimal avian migration: state, time and predation. J Avian Biol 29:395–404
Klaassen M, Lindström Å (1996) Departure fuel loads in time-minimizing migrating birds can be explained by the energy costs of being heavy. J Theor Biol 183:29–34
Kullberg C, Fransson T, Jakobsson S (1996) Impaired predator evasion in fat blackcaps (*Sylvia atricapilla*). Proc R Soc Lond B 263:1671–1675
Kullberg C, Jakobsson S, Fransson T (2000) High migratory fuel load impair predator evasion in sedge warblers. Auk 117:1034–1038
Kvist A, Lindström Å, Green M, Piersma T, Visser GH (2001) Carrying large fuel loads during sustained bird flight is cheaper than expected. Nature 413:730–732
Lind J (2004) What determines probability of surviving predator attacks in bird migration? The relative importance of vigilance and fuel load. J Theor Biol 231:223–227
Lind J, Cresswell W (2006) Anti-predation behaviour during bird migration; the benefit of studying multiple behavioural dimensions. J Ornithol 147:310–316
Lind J, Fransson T, Jakobsson S, Kullberg C (1999) Reduced take-off ability in robins (*Erithacus rubecula*) due to migratory fuel load. Behav Ecol Sociobiol 46:65–70
Lindhe Norberg UM (2004) Bird flight. Acta Zool Sinica 50:921–935
Lindström Å, Alerstam T (1992) Optimal fat loads in migrating birds: a test of the time-minimization hypothesis. Am Nat 140:477–491
Loria DE, Moore FR (1990) Energy demands of migration on red-eyed vireos, *Vireo olivaceus*. Behav Ecol 1:24–35
Moore FR (1994) Resumption of feeding under risk of predation: effect of migratory condition. Anim Behav 48:975–977
Norberg UM (1990) Vertebrate flight. Mechanics, physiology, morphology, ecology and evolution. Springer, Berlin
Norberg UM (1996) Energetics of flight. In: Carey C (ed) Avian energetics and nutritional ecology. Chapman & Hall, NY
Payevsky VA (1985) Demografiya ptits (Avian demography). Nauka, Leningrad
Pennycuick CJ (1975) Mechanics of flight. In: Farner DS, King JR, Parkes KC (eds) Avian biology, vol 5. Academic Press, NY
Pennycuick CJ (1989) Bird flight performance: a practical manual. Oxford University Press, NY
Rayner JMV (1990) The mechanics of flight and bird migration performance. In: Gwinner E (ed) Bird migration. Springer, Berlin

Robinson WD, Bowlin MS, Bisson I, Shamoun-Baranes J, Thorup K, Diehl R, Kunz TH, Mabey S, Winkler DW (2010) Integrating concepts and technologies to advance the study of bird migration. Front Ecol Environ 8:354–361

Salewski V (1999) Untersuchungen zur Überwinterungsökologie paläarktischer Singvögel in Westafrika unter besonderer Berücksichtigung der Wechselwirkungen zu residenten Arten. W&T Verlag, Berlin

Salewski V, Jones P (2006) Palearctic passerines in Afrotropical environments: a review. J Ornithol 147:192–201

Schmidt-Nielsen K (1997) Animal physiology: adaptation and environment, 5th edn. Cambridge University Press, NY

Schmidt-Wellenburg CA, Biebach H, Daan S, Visser GH (2007) Energy expenditure and wing beat frequency in relation to body mass in free flying Barn Swallows (*Hirundo rustica*). J Comp Physiol B 177:327–337

Schmidt-Wellenburg CA, Engel S, Visser GH (2008) Energy expenditure during flight in relation to body mass: effects of natural increases in mass and artificial load in rose coloured starlings. J Comp Physiol B 178:767–777

Stutchbury BJM, Tarof SA, Done T, Gow E, Kramer PM, Tautin J, Fox JW, Afanasyev V (2009) Tracking long-distance songbird migration by using geolocators. Science 323:896

Stutchbury BJM, Gow E, Done T, MacPherson M, Fox JW, Afanasyev V (2011) Effects of post-breeding moult and energetic condition on timing of songbird migration into the tropics. Proc R Soc Lond B 278:131–137

Tobalske BW, Hedrick TL, Dial KP, Biewener AA (2003) Comparative power curves in bird flight. Nature 421:363–366

Torre-Bueno JR, LaRochelle J (1978) The metabolic cost of flight in unrestrained birds. J Exp Biol 75:223–229

Tucker VA (1968) Respiratory exchange and evaporative water loss in the flying budgerigar. J Exp Biol 48:67–87

Tucker VA (1972) Metabolism during flight in the laughing gull *Larus atricilla*. Am J Physiol 222:237–245

Videler JJ (2005) Avian flight. Oxford University Press, Oxford

von Stünzner-Karbe D (1996) Territorialität, Habitatnutzung und Furagierverhalten überwinternder Trauerschnäpper (*Ficedula hypoleuca*) in West-Afrika. Diplomarbeit, Universität Bayreuth

Vrugt JA, van Belle J, Bouten W (2007) Pareto front analysis of flight time and energy use in long-distance bird migration. J Avian Biol 38:432–442

Ward S, Bishop CM, Woakes AJ, Butler PJ (2002) Heart rate and the rate of oxygen consumption of flying and walking barnacle geese (*Branta leucopis*) and bar-headed geese (*Anser indicus*). J Exp Biol 205:3347–3356

Ward S, Möller U, Rayner JMV, Jackson DM, Nachtigall W, Speakman JR (2004) Metabolic power of European starlings *Sturnus vulgaris* during flight in a wind tunnel, estimated from heat transfer modelling, doubly labelled water and mask respirometry. J Exp Biol 207:4291–4298

Weber TP, Houston AI (1997a) Flight costs, flight range and the stopover ecology of migrating birds. J Anim Ecol 66:297–306

Weber TP, Houston AI (1997b) A general model for time-minimising avian migration. J Theor Biol 185:447–458

Weber TP, Houston AI, Ens BJ (1994) Optimal departure fat loads and stopover site use in avian migration: an analytical model. Proc R Soc Lond B 258:29–34

Chapter 5
Habitat Selection and Use by Passerine Migrants

Abstract In this chapter I emphasize that qualitative stopover ecology should be studied in a close association with the stopover behaviour, primarily with habitat use and selection by migrants. During stopovers, migrating songbirds are confronted with unfamiliar habitats, whereas they need to forage efficiently and refuel quickly. The habitat use of migrants is non-random, and I analyse the process of habitat selection and use on the basis of capture data and of radio tracking results of passerine migrants at stopovers. I discuss hierarchical decision making process when ceasing migratory flights and selecting stopover habitat, and whether this process is age-related. The importance of broader landscape context for stopovers, apart from the immediate habitat structure, is emphasized.

5.1 Introductory Remarks

Success of migratory stopover should actually be measured as survival rate during stopover. However, as daily survival rates are difficult to measure and they are anyway very close to 1, fuel deposition rate, or more directly food intake rate may be used as a proxy (under the condition that the migrant survives the stopover). These parameters are not identical for all migrants, but vary with sex and age (Wang et al. 1998; Woodrey 2000; Heise and Moore 2003; Yosef and Chernetsov 2004, 2005) and individually (Chernetsov et al. 2004b). Actual FDR that is reached at a particular site is a result of interactions between an individual migrant and stopover site characteristics, of which habitat quality is one of the most important ones. FDR and the pattern of its variation (local, global, or stochastic) is one of the crucial factors governing stopover duration and the whole organisation of migration (Hedenström and Alerstam 1997; Weber and Houston 1997b; Houston 1998; Chernetsov et al. 2004b). Optimal habitat selection and its subsequent exploitation are most important for successful migration.

N. Chernetsov, *Passerine Migration*, DOI: 10.1007/978-3-642-29020-6_5,

5.2 Scales of Habitat Selection at Stopover

When a landbird migrant makes a stopover, it usually finds itself in unfamiliar surroundings, with food demands and competition not infrequently high (Hutto 1985b; Loria and Moore 1990; Moore and Wang 1991; Chernetsov 2006). Therefore, selection of optimal habitats is of great importance for refuelling migrants. Many migrants which breed at temperate and/or high latitudes and winter in the tropics, have to be able to select the appropriate habitat in boreal forests, steppes, highlands, deserts and tropical areas. Even birds that breed and winter within temperate areas may be challenged. A European robin that breeds in boreal forests of Scandinavia or northern Russia may spend its winter in savannah-like cork oak stands in Portugal or in the cliffs on the Black Sea coast, i.e. in habitats very much unlike its preferred nesting habitat.

It is usually assumed that when selecting optimal stopover habitats, migrants use a hierarchical chain of cues (Hutto 1985b; Moore et al. 2005; Buler et al. 2007; Deppe and Rotenberry 2008). It is also assumed that because migrants in different condition may have varying stopover aims, they may rely on different cues for their condition-dependent habitat assessment. For example, lean birds may need to refuel urgently, even at the cost of accepting predation risk, whereas their fatter conspecifics may be more interested in good shelter (Moore and Aborn 2000; Dierschke 2003; Wang and Moore 2005). When crossing arid areas, like the Sahara or Central Asian deserts, water may be the crucial resource.

The problem of habitat assessment may be especially acute for songbirds that migrate at night and often solo. Captures in the early morning hours, i.e. immediately after ceasing migratory flights, show that the results of initial habitat (pre)selection that occurs prior to and during landfall are usually very precise (Bairlein 1981, 1983). On the Courish Spit on the southeastern Baltic coast, 45 years of capturing migrants in stationary funnel traps located on the border between pine plantations and sandy dunes resulted in captures of 175 Eurasian reed warblers and 190 sedge warblers. Nine years of captures in mist-nets in an optimal habitat 11 km away from that site resulted in the capture of 8,918 Eurasian reed warblers and 6,748 sedge warblers (Mukhin et al. 2005). Obviously, both species, which are quite common passage migrants in the area, were strongly underrepresented in an atypical habitat, which implies that they rarely landed in it. However, accuracy of habitat recognition might be decreased, e.g. under the conditions of poor visibility, like rain of fog (Jenni 1996).

Habitat selection seems to occur during the daylight during and after landfall and is probably mainly based on visual cues. However, birds that land in twilight, i.e. in poor visibility, may also rely on visual cues. It has been reported that Eurasian reed warblers that were landing after a nocturnal flight have mistaken maize stands (a vertically structured habitat) for their optimal habitat, reedbed. This mistake was corrected in the daytime by movements to the optimal habitat (Degen and Jenni 1990). It should be however noted that plum aphids, the preferred food of *Acrocephalus* warblers, may occur in maize, and these birds are known to occur and to forage in maize fields (John Walder, "personal communication"). Therefore, landfall into a maize field may have been a deliberate decision by the birds and not a result of imperfect habitat assessment.

The hierarchical chain of events during habitat selection looks as follows: (1) habitat (pre)selection when still airborne; (2) landfall; (3) redistribution across (micro)habitats, or search; (4) settling; and (5) exploitation of the home range where the migrant has settled. The first link in this chain, preselection when still in flight, is often omitted, but it may be actually an important event that, if performed correctly, greatly simplifies the rest. Cochran et al. (2008) found a specific wingbeat pattern in two species of North American *Catharus* thrushes and in the wood thrush during what they called the final descent phase of nocturnal migratory flight. Wingbeat frequency and the percentage of pauses increased, suggesting slower ground speed, possibly to have a better look at the landscape beneath the birds.

All these events are most probably condition-dependent: individuals with different level of fuel stores may have different objectives during stopovers and different demands for the habitats. During flights across the eastern Sahara lean woodland Palaearctic migrants (e.g. willow warblers) were mainly found in the oases, whereas fatter individuals occurred across a wide range of habitats in the desert (Biebach et al. 1986; Biebach 1990). The same pattern was reported from the western Sahara (Bairlein 1985). It seems that fatter birds were not necessarily trying to refuel, but were just waiting for the night and cooler air temperatures to continue migration. The same reason could explain why in Central Asia most songbird migrants captured in oases were lean, whereas birds trapped in the desert were on average fatter (Dolnik 1990).

Migrants with large fuel stores may be not trying to further increase them, but to emphasize their safety and thus be inactive. They should prefer safer habitats, not necessarily proving good foraging opportunities. Leaner individuals that need to refuel may be more interested in occupying a temporary home range in an optimal habitat that would allow them to reach a high FDR (Chernetsov 2005; Chernetsov and Mukhin 2006). On the other hand, very lean, emaciated migrants may be in dire need to forage immediately and use any opportunity to refuel, without being too finicky about choosing the habitat.

5.3 Role of Individual Experience

Waterfowl and waders that often migrate with discrete stopover opportunity (i.e. they have a limited number of suitable stopover places along their migratory routes) apparently often use the same traditional stopovers year after year (Pienkowski 1976; Evans and Townsend 1988; Pfister et al. 1998; Fox et al. 2002). Whether passerines that usually enjoy continuous stopover opportunity do the same, is an object of discussion.

Long-term trapping project at the Biological Station Rybachy on the Courish Spit during 55 years (1957–2011) did not result in evidence of individual fidelity to stopover sites, in spite of large trapping figures (tens of thousands of individuals trapped annually, a total exceeding 2.7 million captures; Bolshakov et al. 2001, 2002, 2011). It is however worth noting the trapped individuals comprise just a fraction of birds that annually fly over the Courish Spit (Bolshakov 1981; Dolnik et al. 1981). It should be emphasized that flying over the same part of migratory route, known e.g. for Indian sparrows *Passer indicus* and Spanish sparrows *P. hispaniolensis* on Chokpak pass in western Tien Shan Mountains (Gavrilov 1998), is a matter of navigation mechanisms used on migration (Mouritsen 2003); we focus here on repeated using of the same stopover sites.

The often-cited studies that claimed that a substantial proportion of migrants repeatedly used the same stopover sites were performed in Spain (Cantos and Tellería 1994) and in Israel (Merom et al. 2000). The latter study reported a high stopover site fidelity in Eurasian reed warblers (22%, 27 birds out of 123) which is not significantly different from the breeding site fidelity of the same species at the same site (27%, 210 out of 773; $\chi^2 = 1.48$, $p = 0.22$). However, these authors excluded all birds that were only captured in one year which is an unorthodox assumption that might need re-evaluation, as stressed by Catry et al. (2004). A detailed and careful analysis of ringing data from Portugal performed by the authors of the latter paper showed that recurrence of songbirds at their previous year stopover sites was low and did not exceed the values that should be expected from random occurrence within the species-specific optimal habitats. Interestingly enough, birds of wetland habitats that occur sporadically in the Mediterranean region did not show a much higher stopover site fidelity than habitat generalists like e.g. willow warbler and pied flycatcher (Catry et al. 2004). Another report of stopover site fidelity in the bluethroat (Panov 2008) comes from the northern boreal forest of Karelia where these migrants occur in habitat islands in the sub-optimal taiga matrix (Panov 2012).

Generally, it may be concluded that the role of previous experience with a particular stopover site is mostly negligible even in experienced migrants.

5.4 Termination of Migratory Flights

Very few field data is available on when and how nocturnal passerine migrants complete their flights. It is commonly accepted that songbirds that are not crossing ecological barriers take off within a narrow time window after sunset, fly during several hours and cease their flight soon after midnight, i.e. in the darkness long before sunrise (Moore 1987; Kerlinger and Moore 1989). This idea is based on radar data; its critique and new data are given in Sect. 7.1. On the basis of this concept, and because nocturnal migrants usually select the landing habitat rather exactly, it has been hypothesised that habitat selection takes place after sunrise, most probably in the early morning (Moore et al. 1993, 1995). It is believed that 'morning flights', known for many species of nocturnal migrants, support this view (Gauthreaux 1978; Bingman 1980; Wiedner et al. 1992; Yaukey 2010). It should be however kept in mind that morning flight may not be mainly habitat-related, but aimed to compensate for orientation errors and wind drift during the preceding nocturnal flights (Bingman 1980; Moore 1990). The question when songbirds terminate their nocturnal flights is discussed in Sect. 7.3.

Rapid selection of high-quality habitat allows migrants to gain access to stopover resources ahead of competitors (Thompson et al. 2003; Mettke-Hoffmann and Gwinner 2004). Individuals that manage to make selection quickly get an advantage. Therefore, natural selection should probably support habitat selection at an early stage, maybe even before landing, i.e. when ending migratory flights. Two possible strategies of habitat assessment can be used: birds either sample the suitability of the habitat on their own (direct sampling) or use cues from other birds that are already present in a particular habitat (cue using according to Mönkkönen et al. 1999). Cue using allows birds to assess habitats from a distance without sampling, thus making the process more rapid.

Nocturnally migrating songbirds are known to respond to playback of conspecific (Herremans 1990b; Mukhin 2004; Alessi et al. 2010) and heterospecific song (Herremans 1990a) by landing close to the playback site. This response forms the basis of the tape-luring method (Herremans 1990a, b; Schaub et al. 1999; Bulyuk et al. 2000; Mukhin 2004; Mukhin et al. 2005, 2008). We made a study which showed that under some circumstances, in particular during landing before sunrise, acoustic distant cues may be used and prevail over visual stimuli. We have also shown that distant sampling on the basis of acoustic cues is more typical of habitat specialists, wetland specialists in particular, and that during autumn passage adults use acoustic information more actively than hatching-year first-time migrants (Mukhin et al. 2008).

We analysed capture data from two sites on the Courish Spit, at an optimal habitat on Cape Rossitten and at a suboptimal habitat in sandy dunes covered by willow scrub and compared them with historic data on birds killed at lighthouses in Denmark during nocturnal migratory flights in 1886–1939 (Hansen 1954). We tape-lured songbird migrants at a specially selected playback site situated in the transition gap between pine plantations and high sand dunes partly covered with willow scrub (Mukhin et al. 2008). The nearest wetlands were located 4 km to the southwest and 10 km to the northeast of the study site. We considered the dunes where the song playback experiment was performed as a suboptimal habitat for *Acrocephalus* species whose songs we played, because only 175 Eurasian reed warblers and 190 sedge warblers had been captured in stationary funnel traps located 150 m off over a 45 year period (see Sect. 5.2). These long-term trapping data indicate that this site is rarely used by wetland birds, in particular by *Acrocephalus* warblers, under natural conditions.

At night the songs of several passerine species (Eurasian reed warbler, marsh warbler *Acrocephalus palustris*, sedge warbler, pied flycatcher) were played by two car tape players with 30 W loudspeakers. After sunset, we checked the mist-nets on an hourly basis throughout the night. The birds captured earlier than two

Table 5.1 Numbers of birds tape-lured in the suboptimal habitat during the deep night in autumn on the Courish Spit as compared with the numbers killed at lighthouses in Denmark (from the nocturnal flow of migrants)

Species	Tape-luring		Lighthouses, night		χ^2	p
	n	%	n	%		
Great reed warbler	3	0.84	1	0.02	21.5	<0.0001*
Marsh warbler	5	1.40	7	0.13	19.9	<0.0001*
Sedge warbler	27	7.56	142	2.66	26.2	<0.0001*
Eurasian reed warbler	224	62.75	62	1.16	26.45	<0.0001*
Tree pipit	0	0.00	135	2.53	8.2	0.004**
Bluethroat	2	0.56	9	0.17	1.02	0.31
Icterine warbler	0	0.00	68	1.28	3.59	0.058
Red-backed shrike	1	0.28	16	0.30	0.19	0.66
Grasshopper warbler	**2**	**0.56**	**16**	**0.30**	**0.13**	**0.72**
Pied flycatcher	0	0.00	751	14.08	56.7	<0.0001**
Spotted flycatcher	0	0.00	17	0.32	0.32	0.57
Northern wheatear	0	0.00	765	14.35	57.9	<0.0001**
Common redstart	1	0.28	1178	22.09	95.6	<0.0001**
Willow warbler	0	0.00	757	14.20	57.2	<0.0001**
Whinchat	2	0.56	50	0.94	0.19	0.66
Blackcap	22	6.16	267	5.01	0.70	0.40
Garden warbler	61	17.09	854	16.02	0.21	0.65
Whitethroat	4	1.12	163	3.06	3.75	0.053
Lesser whitethroat	2	0.56	59	1.11	0.50	0.48
Barred warbler	1	0.28	15	0.28	0.26	0.61

*The proportion in tape-luring captures is significantly higher than in the flow of migrants
**The proportion in tape-luring captures is significantly lower
Habitat specialists are marked bold

hours before sunrise were considered to be trapped during the deep night, and those captured less than two hours before sunrise, at dawn.

Tape-luring data from a suboptimal habitat were compared with captures in reeds and scrub on the coast of the Courish Lagoon 11 km from site 1. At site 2 which was located in the habitat optimal for *Acrocephalus* warblers and other long-distance passerine migrants, captures occurred without playback. A total of 2,607 birds of 42 species were tape-lured; however, only 17 species were captured during the deep night, before twilight. These birds were with certainty diverted from the flow of migrants aloft by song playback, and not captured during morning movement. The bulk of birds was formed by the Eurasian reed warbler, sedge warbler, garden warbler, blackcap and grasshopper warbler (Tables 5.1–5.4; however, the latter species was only common in twilight captures in autumn). The proportions of other species were low. It is worth noting that such wetland non-passerines as one water rail *Rallus aquaticus*, two spotted crakes *Porzana porzana* and one common sandpiper *Actitis hypoleucos* were tape-lured attracted by Eurasian reed warbler song.

Table 5.2 Numbers of birds tape-lured in the suboptimal habitat in the morning twilight in autumn on the Courish Spit as compared with the numbers captured during daytime in the optimal habitat without song playback

Species	Tape-luring		Optimal habitat, day		χ^2	p
	n	%	n	%		n
Great reed warbler	3	0.47	1	0.38	0.14	0.71
Marsh warbler	10	1.56	10	3.82	3.38	0.066
Sedge warbler	147	22.97	33	12.60	12.5	0.0004*
Eurasian reed warbler	242	37.81	61	23.28	17.6	<0.0001*
Tree pipit	2	0.31	1	0.38	0.22	0.64
Bluethroat	5	0.78	5	1.91	1.35	0.26
Icterine warbler	1	0.16	1	0.38	0.02	0.90
Red-backed shrike	1	0.16	0	0.00	0.21	0.64
Grasshopper warbler	59	9.22	7	2.67	10.8	0.01*
Pied flycatcher	4	0.63	7	2.67	4.88	0.027
Spotted flycatcher	1	0.16	6	2.29	8.4	0.038**
Northern wheatear	3	0.47	0	0.00	0.22	0.64
Common redstart	8	1.25	17	6.49	17.04	<0.0001**
Willow warbler	7	1.09	29	11.07	45.7	<0.0001**
Whinchat	8	1.25	1	0.38	0.68	0.41
Blackcap	49	7.66	55	20.99	31.1	<0.0001**
Garden warbler	81	12.66	23	8.78	2.74	0.098
Whitethroat	6	0.94	0	0.00	1.26	0.26
Lesser whitethroat	3	0.47	5	1.91	2.9	0.089
Barred warbler	0	0.00	0	0.00	–	–

*The proportion in tape-luring captures is significantly higher than in the optimal habitat
**The proportion in tape-luring captures is significantly lower
Habitat specialists are marked bold

In both migratory seasons the vast majority of captures during the night consisted of the Eurasian reed warbler (63% of captured in spring and 91% in autumn), whereas the proportion of this species in the flow of migrants was very low (Tables 5.1, 5.3). In spring other *Acrocephalus* warblers were also overrepresented in song playback captures as compared with the flow of migrants (Table 5.3). At the same time, the common redstart *Phoenicurus phoenicurus*, northern wheatear, pied flycatcher and some other songbirds were underrepresented in song playback captures (Tables 5.1, 5.3). This suggests that Eurasian reed warblers and to some extent their congeners were indeed lured by playing back the Eurasian warbler song to the sandy dunes where they normally do not occur (Mukhin et al. 2005, 2008). The same follows from comparisons of morning tape-luring captures in the sand dunes (site 1) and captures in the optimal habitat without song playback (site 2): in the former case, *Acrocephalus* warblers were more common (Tables 5.2, 5.4). Our data agree with the findings of Alessi et al. (2010) that yellow-breasted chats *Icteria virens* are also lured into unsuitable habitat by broadcast of their song.

Table 5.3 Numbers of birds tape-lured in the suboptimal habitat during the deep night in spring on the Courish Spit as compared with the numbers killed at lighthouses in Denmark (from the nocturnal flow of migrants)

Species	Tape-luring		Lighthouses. night		χ^2	p
	n	%	n	%		n
Great reed warbler	1	0.64	0	0.00	4.14	0.042*
Marsh warbler	1	0.64	1	0.03	1.63	0.20
Sedge warbler	4	2.56	162	5.61	2.1	0.15
Eurasian reed warbler	142	91.03	9	0.31	25.64	<0.0001*
Tree pipit	0	0.00	39	1.35	1.2	0.27
Bluethroat	0	0.00	11	0.38	0.01	0.93
Icterine warbler	0	0.00	5	0.17	0.24	0.62
Red-backed shrike	3	1.92	15	0.52	2.86	0.09
Grasshopper warbler	0	0.00	2	0.07	1.63	0.20
Pied flycatcher	0	0.00	414	14.34	24.7	<0.0001**
Spotted flycatcher	0	0.00	19	0.66	0.24	0.62
Northern wheatear	0	0.00	528	18.28	33.2	<0.0001**
Common redstart	1	0.64	523	18.11	30.5	<0.0001**
Willow warbler	1	0.64	734	25.42	48.2	<0.0001**
Whinchat	0	0.00	177	6.13	9.06	0.003**
Blackcap	0	0.00	31	1.07	0.79	0.37
Garden warbler	0	0.00	36	1.25	1.05	0.31
Whitethroat	1	0.64	135	4.67	4.74	0.029**
Lesser whitethroat	1	0.64	44	1.52	0.3	0.58
Barred warbler	1	0.64	3	0.10	0.45	0.50

*The proportion in tape-luring captures is significantly higher than in the flow of migrants
**The proportion in tape-luring captures is significantly lower
Habitat specialists are marked bold

Our data suggest an important role of acoustic cues in habitat recognition during both spring and autumn migration. It is noteworthy that in autumn, adult Eurasian reed warblers were more attracted by song playback than first-autumn birds. In 1999–2001, after 1 August adults made 15.6% of tape-luring captures (n = 546) and just 6.0% of captures at site 2 without song playback (n = 801; $\chi^2 = 33.4$, $p < 0.001$). The reason for this age-related difference might be that adults, unlike juveniles, have heard species-specific song and have better experience of using acoustic information in different contexts. On the other hand, in coastal areas (like the Courish Spit) the age ratio at stopovers may be biased towards juveniles as compared with the numbers aloft (coastal effect; Payevsky 1985, 1998, 2009). An increased proportion of individuals attracted by song playback from the flow of migrants (which includes proportionally more adults than found at stopovers) may cause an increased proportion of adults in tape-luring captures, reported in Eurasian reed warblers (Mukhin et al. 2005) and in bluethroats in northern Karelia (Panov and Chernetsov 2010b).

Response by first-autumn birds to the species-specific song in autumn, when *Acrocephalus* warblers are no longer singing, suggests an endogenous mechanism

Table 5.4 Numbers of birds tape-lured in the suboptimal habitat in the morning twilight in spring on the Courish Spit as compared with the numbers captured during daytime in the optimal habitat without song playback

Species	Tape-luring		Optimal habitat, day		χ^2	p
	n	%	n	%		n
Great reed warbler	4	1.90	1	0.26	2.73	0.0990
Marsh warbler	2	0.95	60	15.35	29.00	<0.0001
Sedge warbler	49	23.33	48	12.28	12.30	0.0004*
Eurasian reed warbler	116	55.24	122	31.20	27.40	<0.0001*
Tree pipit	1	0.48	1	0.26	0.09	0.7700
Bluethroat	1	0.48	1	0.26	0.09	0.7700
Icterine warbler	1	0.48	1	0.26	0.09	0.7700
Red-backed shrike	0	0.00	1	0.26	0.10	0.7500
Grasshopper warbler	6	2.86	15	3.84	0.15	0.7000
Pied flycatcher	2	0.95	15	3.84	3.15	0.7600
Spotted flycatcher	1	0.48	5	1.28	0.26	0.6100
Northern wheatear	0	0.00	1	0.26	0.10	0.7500
Common redstart	0	0.00	21	5.37	10.15	0.0014**
Willow warbler	4	1.90	5	1.28	0.06	0.8000
Whinchat	7	3.33	11	2.81	0.01	0.9200
Blackcap	1	0.48	35	8.95	15.90	0.0001**
Garden warbler	1	0.48	9	2.30	1.78	0.1800
Whitethroat	8	3.81	25	6.39	1.30	0.2500
Lesser whitethroat	6	2.86	14	3.58	0.05	0.8200
Barred warbler	0	0.00	0	0.00	–	–

*The proportion in tape-luring captures is significantly higher than in the optimal habitat
**The proportion in tape-luring captures is significantly lower
Habitat specialists are marked bold

of habitat recognition on the basis of acoustic stimuli. This mechanism that we have experimentally triggered in autumn normally functions during the spring migration. The song of adult males that are the first to arrive in spring (Payevsky 1985) is a marker of optimal habitat. We have hypothesised that this mechanism may be switched on already in wintering areas, where first-winter Eurasian reed warblers may use the singing of African reed warblers *Acrocephalus baeticatus* which is very similar to the song of the Eurasian species (Cramp 1992). A recent study of the process of the future breeding site selection in black-throated blue warblers *Dendroica caerulescens* showed that juveniles responded to the playback of adult song, i.e. use it as a cue to habitat suitability, already during the postfledging movements (Betts et al. 2008). This finding makes the use of acoustic stimuli during the autumn migration by *Acrocephalus* warblers for habitat recognition less astonishing.

Most interesting is the response to heterospecific song. A total of 17 species responded to the Eurasian reed warbler song during the four years of study (Mukhin et al. 2005). Most of these birds were other *Acrocephalus* warblers. The most obvious response was shown by the sedge warbler, both to its own song and to the Eurasian reed warbler song. On the other hand, playing back the redwing *Turdus iliacus* song on 7–25 May 2002 attracted as few as nine individuals: three lesser whitethroats, two red-backed shrikes, a marsh and a reed warbler, a European robin and a whinchat *Saxicola rubetra* (one individual of each species). Only three birds were captured before twilight. Not every bird song is attractive, but only that of species typical of certain habitats (wetlands in particular) which may be acoustic markers of such habitat (Mukhin et al. 2008). Distant markers that make indirect estimates of habitat quality possible may be more relevant for habitat specialists, especially those preferring fragmented habitats. However, birds with more general patterns of habitat use, like North American thrushes, also seem to be using distant cues. In the final phase of nocturnal flights thrushes change their flight direction and increase wingbeat frequency (Bowlin et al. 2005; Cochran et al. 2008), probably to decrease their ground speed. Most probably these (and maybe other) migrants visually explore the landscape and choose the site to land.

5.5 Search for Home Range and Settling

5.5.1 Range and Pattern of Morning Movements

After landing, migrants sample habitat quality at their new stopover site. Movements of newly grounded migrants often result in the morning peak of captures in standardised trapping projects, familiar to every bird ringer (Dolnik and Yablonkevich 1985; Brensing 1989; Berthold et al. 1991; Titov 1999b).

It should be emphasized that 'morning flights' of nocturnal migrants (see the previous section) are not the morning peak of captures. Morning flights occur not everywhere and not all the time. They occur above the canopy and always higher than standard mist-nets are normally put up (usually their upper edge is 2–2.5 m above the ground). Birds that participate in morning flight are usually not captured in mist-nets. These flights most probably are performed to compensate wind drift or to move inland from the coast (Gauthreaux 1978; Bingman 1980), not to select optimal habitat. Conversely, the morning peak of movements within the canopy (and thus of captures) does suggest that fine-tuning of (micro)habitat selection in many passerine nocturnal migrants occurs during several hours after sunrise.

Some nocturnal migrants, e.g. European robins, perform longer movements (for 1–2 km), mainly within several hours after sunrise (Chernetsov 2005; Chernetsov and Mukhin 2006; Tsvey 2008). During the first day of stopover in spring and the first two days in autumn, radio-tagged European robins on the Courish Spit moved more broadly than in subsequent days, as shown by the daily values of the linearity index of their movements (Chernetsov and Mukhin 2006). Summer tanagers behave in a similar manner (Aborn and Moore 1997), and several other North American nocturnal migrants may also cover up to 2.0–2.5 km before they settle

(Taylor et al. 2011). However, other species, e.g. pied flycatchers, may perform relatively long movements (for several kilometres) during any day of stopover, also several days after their arrival (Chernetsov et al. 2004a). Such relatively long-distance movements of pied flycatchers and possibly other species should not be regarded as extended search. It seems more correct to say that in some species the exploitation of resources at stopover is based on broader movements than in others.

5.5.2 Search and Settling Time

As already mentioned, search and settling time varies between several hours (some individuals may occupy their home range very quickly, within minutes) and one, rarely two, days. European robins do not start to gain mass until they occupy a limited home range (Titov 1999a). The search/settling period apparently is usually characterised by negative FDR. Therefore time which a migrant is ready to invest into search and settling is an important parameter that influences its optimal migration strategy (Weber and Houston 1997a; Chernetsov et al. 2004b). Its value most probably depends on fuel stores at arrival (Chernetsov et al. 2004b). Time that a migrant is ready to spend exploring the stopover area should depend on temporal and energetic costs of exploration and on potential benefits that this information may bring (Aborn and Moore 1997). Both costs and benefits are likely condition-dependent. One can imagine that individuals with relatively large fuel stores may venture to be choosy, and maybe even to skip a site that cannot provide them with sufficiently high FDR and to depart during the next night. Continuing our metaphor of a human job seeker, someone with large personal means may choose to remain unemployed and continue looking for a perfect job for a longer time than someone who has bills that must be urgently paid. On the other hand, fatter birds may decide to continue flight in the next night anyway and not invest into looking for a good stopover site. In this case they should conserve energy and remain stationary. This was what fat Eurasian reed warblers tape-lured into a poor habitat on the Courish Spit did (Ktitorov et al. 2010).

Our radio-tagging study of stopover duration and stopover spatial behaviour of fat and lean European robins on the Courish Spit did not show any clear relationship between arrival fuel load and stopover behaviour (Chernetsov and Mukhin 2006). The area of the individual home range did not differ between robins that arrived fat ($n = 5$) and that arrived lean ($n = 8$; t-test, $t = 0.44$; $p = 0.67$). In the first two days of stopover, linearity of movements was not condition-related, either. It is especially interesting that the linearity index of movements on the first day upon arrival did not differ between the birds that departed after the first day and those that stopped over for a longer period (Chernetsov and Mukhin 2006). This means that the pattern of movements of a freshly arrived European robin does not allow prediction whether it is going to depart or to stay, i.e. whether it is a flyer or a feeder.

Table 5.5 Proportions of first-autumn birds among song thrushes and blackbirds captured on the Courish Spit in autumn (stationary funnel traps, passive and active trapping in mist-nest) and in the flow of nocturnal migrants in Denmark (Hansen 1954)

Species	Proportion of juveniles, %			
	Nocturnal flow	Stationary traps	Passive mist-netting	Active mist-netting
Song thrush	37.4	6.5 ± 0.7	3.5 (1.5–4.4 in various years)	15.7 (n = 51)
Blackbird			♂♂13.6; ♀♀ 15.1 (7.7–18.3 in various years)	50 (n = 30)

5.5.3 *Coastal Effect: Are Movements of Migrants Age-Related?*

Mobility of migrants immediately after landing may also depend on the age of birds. This is suggested by the analysis of the coastal effect, which is the abnormally high proportion of juveniles (first-autumn birds in autumn and second-year individuals in spring) among nocturnal migrants captured on the coasts of large water bodies (Ralph 1978, 1981; Dunn and Nol 1980; Payevsky 1998, 2009). The causes of this phenomenon remain unclear. It is usually assumed that the coastal effect is caused by differential behaviour of adults and juveniles when ceasing flight over large water bodies: juveniles go to the nearest land, whereas adults more often continue further inland where they may expect better-quality habitats (Payevsky 1985, 2009).

Our trapping data on song thrushes and blackbirds *Turdus merula* at stopovers on the Courish Spit suggest that in these species the coastal effect may be partly explained by differential mobility of adults and juveniles. When the birds were flushed into the nets (so-called 'active trapping') the proportion of adults in capture was higher than during the routine 'passive' trapping (Table 5.5). One may expect that when stopover migrants are flushed into the nets, adults and juveniles should be captured proportionally to their occurrence in the vicinity of the nets, or at least that age-related bias should be substantially lower than when captures result from spontaneous movements only. The reason for the increased mobility of juveniles might be their subordinate social status (Gauthreaux 1978) which has also been recorded at stopovers (Woodrey 2000). Because of that juveniles may spend more time and move more before they finally manage to occupy their own home range.

It is however worth noting that during passive trapping of migrating song thrushes in the Dzhanybek oasis in the semi-desert of western Kazakhstan (an inland site), the proportion of adults in autumn 2004 was 18.4% (n = 103), i.e. it was very similar to the value found during active trapping on the Courish Spit (coastal site; Table 5.5: Yates-corrected $\chi^2 = 0.04$, $p = 0.84$). We did not do active trapping in Dzhanybek, but a significantly higher proportion of adult song thrushes there strongly suggests that age-related mobility of stopover migrants may be only a partial explanation of the coastal effect.

In the European robin on the Courish Spit, proportions of adults did not differ between active and passive trapping in mist-nets (spring: 14.3 and 13.3%, $n = 233$ and $n = 389$, respectively; $\chi^2 = 0.14$, $p = 0.70$; autumn: 5.1 and 5.0%, $n = 156$ and $n = 685$, respectively; $\chi^2 = 0.01$, $p = 0.93$). For comparison, at autumn stopovers in Dzhanybek the proportion of adults during passive trapping was 25.0% ($n = 280$). Thus, for the European robins we have no reason to assume the age-related level of mobility during search and settling at stopover. It may be mentioned that social status of migrants does not always depend on age, e.g. red-eyed vireos showed no age-related difference in status (Moore et al. 2003).

An explanation of coastal effect that does not involve differential mobility is age-related migration strategy. We have shown that first-autumn bluethroats in northern Karelia have lower FDR than the adults and make longer stopovers more often, whereas adults frequently stop over for one day only (Panov and Chernetsov 2010a). Adult bluethroats (and possibly other nocturnal migrants) may be underrepresented in captures because of their brief stopovers, whereas their proportion among migrants aloft may be significantly higher (e.g. Mukhin et al. 2005), representative of their actual population productivity. If and when age-related variation in the length of migratory flights and stopover duration is smaller (e.g. at some inland sites, for instance in western Kazakhstan), the coastal effect may be absent.

5.5.4 Broad Movements Throughout Stopover

Within species that move long distances (at the scales of several kilometres) during their migratory stopovers it probably makes little sense to speak of the search/settling period, for the simple reason that they never settle. Species that need to occupy a limited home range to start gaining mass, usually either settle or resume migration. Another possible option would be to perform a landscape-scale nocturnal flight for a distance of 5–30 km, as recently reported for Swainson's thrushes, hermit thrushes, black-throated blue warblers and white-throated sparrows *Zonotrichia albicollis* (Mills et al. 2011; Taylor et al. 2011). However, some species may exploit stopover resources when making relatively broad-scale daytime movements, as found in e.g. pied flycatchers on the Courish Spit (Chernetsov et al. 2004a), garden warblers (Fransson et al. 2008) and probably other species (Taylor et al. 2011). The overall distance that pied flycatchers covered during their daytime movements reached 4.0 km (Chernetsov et al. 2004a), and it should be emphasized that (1) this distance was not covered on the first day upon arrival; and (2) this is a minimum estimate, because the bird in question was lost when moving through dense forest. In reality they may have covered even longer distance (Table 5.6, Fig. 5.1). Garden warblers on Crete moved up to 5 km during stopover (Fransson et al. 2008).

This behaviour contrasts sharply with the behaviour of European robins that usually remain within a much more limited area (Fig. 5.1). Such a strategy of habitat exploitation by the pied flycatcher (and possibly by some other species) may result from relatively low costs of exploratory behaviour and/or high benefits

Table 5.6 Parameters of diurnal movements of radio-tagged pied flycatchers during their stopovers on the Courish Spit. For the individuals lost during their movements, minimum estimates are given

#	Tagging date	Tagging time	Stopover duration,days	Result of tracking	Location number	Overall distance moved (km)	Distance moved daily (km)	Linearity index of movements
006	28.04	7:50	≥ 1	Lost	8	3.80	3.80	0.910
013	29.04	9:50	≥ 3	Lost	20	4.00	3.04; 0.84; 1.02	0.589
021	30.04	14:50	2	Nocturnal departure	10	3.10	0.13; 3.10	0.730
027	02.05	11:50	1	Nocturnal departure	34	0.41	0.41	0.077
037	02.05	15:50	≥ 4	Lost	47	2.30	0.19; 0.17; 0.06; 2.30	0.108
041	08.05	14:10	1	Nocturnal departure	14	0.41	0.41	0.061
059	12.05	12:10	1	Nocturnal departure	8	0.27	0.27	0.474

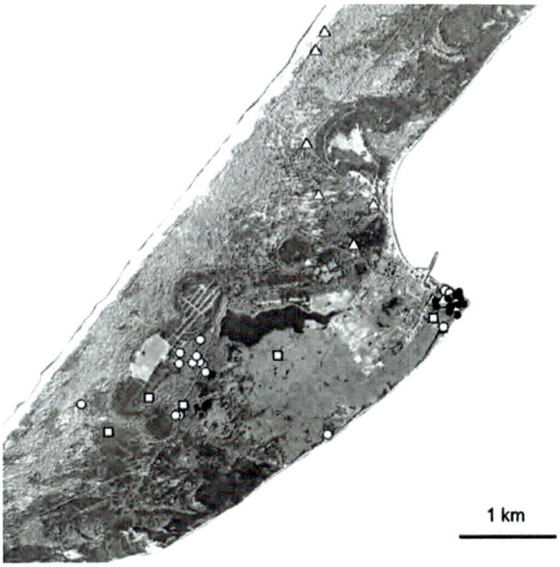

Fig. 5.1 The range of movements of pied flycatchers (*white* symbols) and European robins (*black dots*) during spring migratory stopovers. Different *white symbols* represent different pied flycatcher individuals in May 2003; black dots refer to the single European robin in April 2003. The stopover duration was 1–4 days in pied flycatchers and 3 days in the European robin

for exploring new areas, which is probably the case on the Courish Spit in spring. It is worth noting that during autumn stopovers in Portugal pied flycatchers reportedly occupy temporary territories, i.e. use a totally different strategy (Bibby and Green 1980). This contrasting stopover behaviour by the same species is most probably related to exploitation of different habitats, i.e. habitat dictates spatial strategy.

5.6 Habitat Exploitation

5.6.1 Non-Random Habitat Use

When migrants occupy a home range (if their spatial strategy includes this option), or during their broad movements across habitat, they apparently select and use habitats in a non-random manner (Bairlein 1981, 1983, 1992; Hutto 1985a, b; Ormerod 1990; Moore et al. 1990, 1993). As shown by capture results in different years at a standardised trapping site where the nets are located in exactly the same places year after year, the species-specific patterns of occurrence across habitats remain constant, even though the site is visited each year by different individuals (Bairlein 1981). This shows that the observed pattern is non-random and is typical of each species. First-autumn individuals usually utilise a broader spectrum of habitats than adults, which is usually explained by their less perfect ability to select the optimal habitat (Bairlein 1981, 1983). Another possible explanation of age-related difference in habitat use is that juvenile birds are displaced by adults into suboptimal habitats. Some songbird species alter their habitat preferences during the migratory period as compared to the breeding season. For instance, sedge warblers that usually do not breed in reedbeds, apparently prefer this habitat during their postbreeding and postfledging movements and during autumn and spring migration (Chernetsov 1998). Eurasian reed warblers redistribute within the reed stands: during the period of migration they start using loose stands that provide food but are poorly suitable for building nests (Chernetsov 1998). Migrating European robins and blue tits *Cyanistes caeruleus* during autumn passage spend much time and actively forage in reedbeds that are not used by these species during their breeding season.

The degree of habitat selectively varies between the species. Parnell (1969) found that some New World warblers used a broader array of habitats than others. Similarly, different European songbird migrants vary in their habitat selectivity (Bairlein 1983, Berthold et al. 1991). Whereas European robins occur within a very broad range of habitats, from different forest types through willow scrub on sandy dunes (Tsvey 2008) and even to reedbeds (see above), e.g. Eurasian reed warblers and sedge warblers normally only occur in reed stands during migration, and when tape-lured into other habitats invariably depart at the first opportunity (i.e. during the first night; Ktitorov et al. 2010).

5.6.2 Factors that Govern Habitat Selection by Migrants

The pattern of habitat exploitation is defined by a combination of several factors: endogenous preferences and functional morphology (Bairlein 1983, 1992), foraging strategy and the spatial distribution of food (Hutto 1985a; Martin and Karr 1986; Chernetsov 1998; Titov 2000; Chernetsov and Titov 2003), habitat carrying capacity and density of competitors (Hutto 1985b). Besides, predation risk is another important habitat characteristic which significantly influences the appeal of a habitat to stopover migrants (Alerstam and Lindström 1990; Dierschke 2003; Lank and Ydenberg 2003; Sapir et al. 2004).

These factors deserve special attention. Endogenous habitat preferences based on morphology during migration probably do not differ much from other parts of the annual cycle. At least, no data suggests it. Apparently, species morphologically adapted e.g. to reedbeds with their vertical structure, e.g. Eurasian reed warblers and great reed warblers *Acrocephalus arundinaceus* (Winkler and Leisler 1985; Leisler et al. 1989), will use such habitats also during migration. Moreover, such species have special behavioural adaptations that make it easier for them to locate the patches of this specific and highly fragmented habitat (Sect. 5.4).

It should not be assumed, however, that all migrating songbirds must use the same habitats during migration and wintering that they utilise during the breeding season. I have just mentioned the use of reedbeds by European robins and blue tits. Our stomach flushing data suggest that during autumn passage on the Courish Spit, willow warblers actively feed on plum aphids that they collect in the reedbeds.

Wintering European robins often use habitats that are very much unlike deciduous and coniferous forests where they prefer to breed (Malchevsky and Pukinsky 1983). In the Iberian Peninsula these birds may spend their winter in thin savannah-like cork oak plantations, and in the coastal areas of Abkhazia, where the density of wintering European robins may be very high, they occupy literally all the available habitats, including stony slopes near the Black Sea coast (Chernetsov, unpubl.). Generally speaking, habitat requirements in winter quarters may or may not be similar to the habitat requirements in the breeding range, with habitat specialists showing the strongest correspondence between breeding and winter habitats (Leisler and Schulze-Hagen 2011). Thus, the role of endogenous habitat preferences based on ecomorphological adaptations (that have been experimentally shown to exist, Ley 1988; Grünberger and Leisler 1990, 1993) in habitat selection during migration should not be overestimated.

The pattern of spatial distribution of food resources significantly influences the spatial behaviour of migrants at stopover (see Chap. 6). The same species may utilise different habitats, and even the same habitat, differently: pied flycatchers remain within a limited area and even reportedly occupy territories at stopovers during autumn passage in Portugal (Bibby and Green 1980), but move broadly in spring on the Courish Spit (Fig. 5.1). Rufous hummingbirds *Selasphorus rufus* at stopovers may defend territories, but also may behave as non-territorial roamers depending on their social status (Carpenter et al. 1993b). Apparently, if habitat shows fine-scale variation, the scale of movements will significantly influence habitat use by stopover migrants.

Besides, habitat quality strongly depends on the quality of cover it provides. It has been experimentally shown that blackcaps in migratory disposition started to forage more intensively when they were shown a stuffed sparrowhawk (Fransson and Weber 1997). The authors suggest that blackcaps did it to reach the departure threshold fuel stores sooner and to spend less time at a dangerous site. It is difficult to understand why blackcaps did not try to reach the highest possible FDR anyway (especially if they were time minimising migrants), but irrespectively of interpretations, this study has shown that migrating passerines may adjust their stopover behaviour in respect to predation situation. A field study of behaviour of northern wheatears during autumn stopovers on Helgoland, conversely, showed that migrants significantly decreased their FDR when predation danger (measured as the rate of real raptors flying over) was high (Schmaljohann and Dierschke 2005). The data on relationship between FDR and predation danger is thus equivocal, but a significant influence seems to be a fact. Another thing is that the claim of the authors that 'wheatears do not directly minimise predation risk during migration' should be treated with caution. We have seen in Chap. 4 that the optimal migration theory is very sensitive to *a priori* assumptions that are often made implicitly. By playing with assumptions, one can 'prove' a lot of things in the framework of this theory.

5.6.3 The Importance of Landscape Context

Apart from the immediate local habitat environment, the more general landscape context may play an important role. The response of birds to the general landscape structure is well known for breeding individuals (Hinsley et al. 1995), but in the studies of songbird stopover ecology the importance of landscape context is underestimated (Freemark et al. 1995). It is intuitively clear that when migrating birds are crossing areas with restricted stopover opportunities, like large water bodies or deserts, they should use different criteria of habitat selection than when flying over ecologically hospitable areas. It has been shown that on the Great Plains in North America the density of songbird migrants in spring was inversely proportional to the area of forest patches (Martin 1980). Territoriality of some passerines at stopovers (Rappole and Warner 1976; Kordic-Brown and Brown 1978; Bibby and Green 1980, 1981; Carpenter et al. 1983, 1993a, b; see Chap. 6 for more details) and density-dependent occupation of home ranges (Veiga 1986; Hansson and Pettersson 1989; Moore and Wang 1991; Shochat et al. 2002; Kelly et al. 2002; Ottich and Dierschke 2003) suggest inter- and intraspecific competition between migrants at stopovers. Direct evidence for interference competition is scarce (e.g. Salewski et al. 2007), but the reason is the methodological difficulty to demonstrate competition formally, rather than scarcity of this event in the wild.

When habitats are strongly fragmented, which is commonplace in the areas under much anthropogenic pressure (i.e., sadly, across most of the globe), patch size and the degree of their isolation significantly influence migrant density and thus the severity of competition. It should be expected that FDR will be related to habitat structure. At a stopover in Israel *Sylvia* warblers were less common in an optimal habitat than in a suboptimal one (Shochat et al. 2002). The pattern observed was different from the ideal free distribution. The authors concluded that the observed pattern of distribution of stopover migrants was due to isolation of the suitable habitat patches. The distance between habitat patches was about one kilometre (Shochat et al. 2002).

Our data on energy condition of songbird migrants in different habitats in an oasis in western Kazakhstan suggest that this interpretation should be treated with caution. In autumn 2004 we captured passerines in two habitats in Dzhanybek (49°24′ N, 46°48′ E). One habitat was a plantation of various trees and shrubs, including those fruit-bearing ones: the rowan *Sorbus aucuparia*, single-seeded hawthorn *Crataegus monogyna*, common barberry *Berberis vulgaris*, hedge cotoneaster *Cotoneaster lucidus* etc. (for a more detailed description see Karandina and Erpert 1972). In the diet of garden warblers these berries played a significant role, as shown by inspection of their faeces. Another habitat was the belt of windbreak forest which consisted of Siberian elms *Ulmus pumila*. Even though the former plantation was apparently a rich, and windbreak forest a poor habitat (especially for garden warblers, but also for other species), only two songbirds out of the six analysed showed habitat-related variation in body mass: the common redstart and the chiffchaff *Phylloscopus*

Table 5.7 The habitat-related difference in body mass of migrants captured in Dzhanybek (western Kazakhstan) in autumn 2004

Species	Forest plantation	Windbreak belt	Body mass difference	Wing length difference
Garden warbler	357	68	t = 0.69, p = 0.49	t = 0.08, p = 0.94
Common redstart	288	172	t = 2.02, p = 0.044	t = 0.15, p = 0.88
Chiffchaff	290	25	t = 2.31, p = 0.028	t = 1.73, p = 0.09
Willow warbler	192	31	t = 1.26, p = 0.21	t = 0.09, p = 0.93
European robin	193	18	t = 0.98, p = 0.34	t = 0.57, p = 0.58
Red-breasted flycatcher	483	83	t = 0.27, p = 0.78	t = 1.08, p = 0.28

collybita (Table 5.7). Common redstarts were predictably heavier in the plantation than in windbreak forest belt (15.22 g [SE = 0.099, n = 288] vs. 13.91 g [SE = 0.115, n = 172]). Chiffchaffs, conversely, were lighter in the apparently optimal habitat (7.69 g [SE = 0.051, n = 290] vs. 8.10 g [SE = 0.167, n = 25]), but because of size difference condition index (size-corrected body mass) did not differ between the habitats: t = 1.78, p = 0.086.

We can conclude that the body condition of birds in an apparently poor habitat (windbreak forest belt) was generally not poorer than in an apparently optimal habitat (forest plantation with fruit-carrying shrubs), and the difference was not significant even in such frugivorous migrants as the garden warbler (Table 5.7). This example suggests that before claiming that some habitat is suboptimal for stopover migrants because of food availability, shelter, or physical structure, it is necessary to show by analysing the diet, foraging intensity, FDR etc. that this is indeed the case.

On the other hand, Ktitorov et al. (2008) showed that the proportion of forest habitat within 2–5 km from a trapping site significantly influences FDR of willow warblers and common redstarts during autumn passage throughout Europe. This study showed that landscape context indeed had an impact on stopover site quality in addition to the local habitat situation. Landscape context was shown to influence not the density of stopover migrants, but directly the FDR of those individuals that stopped over at certain sites. FDR, as we have seen in Chap. 3, is one of the principal characteristics of stopover. Buler (2006) who used the density of migrants as a dependent variable, obtained a similar scale of influence of landscape characteristics: 4.9 km.

It should be mentioned that Ktitorov et al. (2008) used a rather simplistic binary classification of habitats: forest habitats were tested against all other types of habitat. The proportion of forest habitats explained a considerable proportion of variation in FDR when forest did not occupy more than 10% of the area in the surroundings. Ten percent was the threshold after exceeding which the proportion of forest habitats in the landscape did not influence FDR of migrants any more (Ktitorov et al. 2008). This means that landscape context is mostly important when the optimal habitat is scarce, i.e. its availability is a limiting factor. This situation is driven to extreme in the case of island habitats and physical offshore islands.

5.7 Fuel Deposition in Oases

Stopovers in small islands and in oases in deserts are an extreme case of contrasting habitat quality between small patches of the optimal (or at least somehow suitable) habitat and very large areas of low-quality matrix. The possibility to select a suitable stopover habitat are strongly limited. One may expect that in such situations low values of FDR should occur more frequently that when migrating over suitable continental areas, because stopover migrants have to accept sites with low FDR: in the matrix habitat, stopover is either completely impossible (water) or there is no chance to achieve a positive FDR (desert). Variation of FDR values recorded in oases should also be higher because of frequent occurrence of low values.

Dolnik (1990) summarised the results of earlier studies of stopover ecology and behaviour of passerines in Central Asia (Dolnik 1982, 1985a, b, 1987). He concluded that among songbirds that stopped over in oases, body mass change was positive in 40% in autumn and in 48% in spring. The proportion of birds gaining mass was inversely related to their mean initial body mass in the given oasis. Generally Dolnik (1990) concluded that small oases in Central Asia did not give small passerines an adequate opportunity to refuel. Because of competition between migrants, the mean FDR in oases in this region was not significantly higher than in open desert.

We studied refuelling in several species of long-distance migrants in a large oasis in Eilat (Israel) on the northernmost tip of the Red Sea (Yosef and Chernetsov 2004, 2005) and in a small oasis in Dzhanybek on the border between Volgograd Region of Russia and Western Kazakhstan Region of Kazakhstan (Chernetsov et al. 2007). In Eilat the FDR of Eurasian reed warblers varied between –1.00 and 1.25 g day^{-1}, on average 0.117 g day^{-1} ± SE = 0.011 (n = 429). The difference between spring values, when the birds were recovering after the flight across the Arabian Desert, and autumn, when they were fuelling up to cross the desert, were non-significant ($t_{230,197} = 0.26$, $p = 0.80$). In the sedge warbler in the same area, FDR varied between −1.50 and 2.40 g day^{-1}, on average 0.140 g day^{-1} ± SE = 0.016 (n = 366). Like in the former species, season-related variation in FDR was not significant ($t_{315,49} = 1.32$, $p = 0.19$). The difference between these two species was not significant, either: $t_{365,428} = 1.16$, $p = 0.25$.

On the Courish Spit which is not located at the edge of an ecological barrier, FDR in the Eurasian reed warbler did not vary between the seasons, either ($t_{747,111} = 1.50$, $p = 0.13$) and was on average 0.037 g day^{-1} ± SE = 0.008 (n = 829). The mean FDR was significantly lower than in Eilat ($t_{858,428} = 5.70$, $p < 0.001$), but dispersions of values did not differ (Fisher's test: $F_{858,428} = 1.07$, $p = 0.21$). In the sedge warbler on the Courish Spit season-related variation was not significant, either ($t_{97,380} = 0.37$, $p = 0.71$), and the mean FDR was 0.065 g day^{-1} ± SE = 0.011 (n = 479). Like the Eurasian reed warbler, the mean FDR in the sedge warbler was significantly lower than in Eilat ($t_{478,365} = 3.81$, $p < 0.001$), but the dispersion of values was significantly lower on the Courish Spit than in Eilat (Fisher's test: $F_{478,365} = 1.62$, $p < 0.001$). The coefficient of variation of FDR values on the

Table 5.8 The body mass change in songbird migrants in Dzhanybek (western Kazakhstan) in autumn

Species	Mean body mass change (g ± SE)	Mean rate of body mass change (g day^{-1} ± SE)	Coefficient of variation of body mass change rate	n	Threshold stopover duration (days)	Threshold initial mass (g)
Garden warbler	0.20 ± 0.202	−0.11 ± 0.065	5.01	72	1.3	20.3
Common redstart	0.16 ± 0.136	0.10 ± 0.082	6.56	64	0	15.7
Chiffchaff	−0.02 ± 0.115	−0.02 ± 0.041	11.04	29	2.6	7.8
Willow warbler	0.08 ± 0.202	0.09 ± 0.107	3.94	11	–	–
European robin	−0.19 ± 0.090	−0.08 ± 0.047	4.24	52	–	–
Red-breasted flycatcher	0.04 ± 0.048	0.03 ± 0.032	11.82	123	–	9.5

In the species which showed a relationship between body mass change and stopover duration and initial body mass, the threshold values of the latter two parameters are given. The birds gain body mass when stopover duration above the threshold and body mass below the threshold

Courish Spit was 6.34 in the Eurasian reed warbler and 3.70 in the sedge warbler. In Eilat the respective values were 1.95 and 2.19.

In a small oasis at the edge of clay semi-desert in Dzhanybek the mean FDR varied between −0.11 g day^{-1} in the garden warbler and 0.10 g day^{-1} in the common redstart (Table 5.8). The coefficient of variation of FDR values varied between 3.94 in the willow warbler and 11.82 in the red-breasted flycatcher. In Rybachy in the same species, except for the red-breasted flycatcher which is rare on the Courish Spit, the respective values ranged from 10.6 in the chiffchaff to 32.5 in the European robin (Table 5.9).

In four species out of five (the only exception was the chiffchaff) the coefficients of variation of FDR were higher in Dzhanybek than on the Courish Spit, and in the European robin the difference was by an order of magnitude. Mass gain rate did not differ between the two sites in four out of five species (t-test, $t > 0.05$), and only in the garden warbler it was significantly different ($t = 1.99$, $p = 0.045$). It should be noted that FDR in this species in Rybachy was close to zero, 0.03 g day^{-1} (SE = 0.024), whereas in Dzhanybek garden warblers were losing mass with the average rate of 0.11 g day^{-1} (SE = 0.065, Tables 5.8, 5.9). However, the mean change of body mass during a stopover in this oasis was positive. The cause of this apparent paradox was that garden warblers that made short stopovers were quickly losing mass, whereas those birds that stopped over for a longer period of time (more than 1.3 days on average, Table 5.8) gained mass (Chernetsov et al. 2007).

During spring migration across the Sahara in several passerine nocturnal migrants in Mauritania body mass and fuel stores of birds that stopped in an oasis were higher than in their conspecifics found in the desert. Moreover, most birds were gaining mass in an oasis in Ouadâne (Salewski et al. 2010b). This data is in contrast with the findings of Biebach et al. (1986) who argued that mainly lean birds were found in oases where they tried to refuel and Dolnik (1990) who added that they mainly failed to do so. However, the data from Mauritania agree with our

Table 5.9 The mean mass change rate in passerine migrants in autumn in Rybachy (Courish Spit)

Вид	Mean rate of body mass change (g day^{-1} ± SE)	Coefficient of variation of body mass change rate	n
Garden warbler	0.03 ± 0.024	14.56	391
Common redstart	0.02 ± 0.038	18.83	76
Chiffchaff	0.02 ± 0.014	10.68	171
Willow warbler	0.03 ± 0.028	12.23	173
European robin	−0.01 ± 0.004	32.54	6365

findings (Yosef and Chernetsov 2004, 2005; Chernetsov et al. 2007) that transient migrants refuel in oases, both large and small ones. FDR in oases did not differ from the sites located in areas with continuous stopover opportunity or was higher at the edge of barriers.

5.8 Fuel Deposition on Islands

Fuel deposition rate of European robins during autumn passage on Helgoland in the North Sea (0.04–0.11 g day^{-1} depending on the progress of the season (Ottich and Dierschke 2003) did not differ from the respective values recorded on the Courish Spit (on average 0.064 g day^{-1}; Tsvey 2008). The mean body mass change of European robins initially captured during a strong fallout of migrants in autumn on the island of Greifswalder Oie in the western Baltic was 0.26 g (n = 175), which together with the mean stopover duration of 2.67 days gives a mean estimate of 0.097 g day^{-1} (Mädlow 1997). This value is by 50% higher than on the Courish Spit. At both sites birds recaptured on the first and second day of stopovers lost mass, and started to regain it since the third day upon arrival (Mädlow 1997; Titov 1999a, c). Fuel deposition rates of garden warblers during autumn migration on Greifswalder Oie and on Helgoland were among the highest values recorded at European sites (Schaub and Jenni 2000).

In conclusion, the hypothesis of low mean FDR in songbird migrants and of high variation of this parameter in oases and on islands is not supported by the data, either from large or from small oases. Apparently, the oases where passerines regularly stop over for more than one day normally offer them adequate refuelling opportunities. At sites where such opportunity is lacking (e.g. in open desert), only one-day stopovers are made (see also Biebach et al. 1986; Salewski et al. 2010). The idea that oases are 'black holes' [or, in current terminology, ecological traps (Schlaepfer et al. 2002)] for emaciated passerine migrants where they stop but cannot refuel (Dolnik 1990) is not supported by the data. It cannot be ruled out that situations of overcrowding do occur in small oases in the centre of large deserts or on small isolated offshore islands, when the small patches of suitable habitat cannot support the large numbers of grounded migrants. However, such situations that likely result in catastrophic mortality of migrating birds are probably uncommon (Payevsky 1999).

References

Aborn DA, Moore FR (1997) Pattern of movement by summer tanagers (*Piranga rubra*) during migratory stopover: a telemetry study. Behaviour 134:1077–1100

Alerstam T, Lindström Å (1990) Optimal bird migration: the relative importance of time, energy, and safety. In: Gwinner E (ed) Bird migration. Springer, Berlin

Alessi MG, Benson TJ, Ward MP (2010) Nocturnal social cues attract migrating yellow-breasted chats. Wilson J Ornithol 122:780–783

Bairlein F (1981) Ökosystemanalyse der Rastplätze von Zugvögeln. Ökol Vögel (Ecol Birds) 3:7–137

Bairlein F (1983) Habitat selection and associations of species in European passerine birds during southward, post-breeding migrations. Ornis Scand 14:239–245

Bairlein F (1985) Body weights and fat deposition of Palaearctic passerine migrants in the central Sahara. Oecologia 66:141–146

Bairlein F (1992) Morphology-habitat relationships in migrating songbirds. In: Hagan JM III, Johnston DW (eds) Ecology and conservation of Neotropical migrant landbirds. Smithsonian Institution Press, Washington

Berthold P, Fliege G, Heine G, Querner U, Schlenker R (1991) Wegzug, Rastverhalten, Biometrie, und Mauser von Kleinvögeln in Mitteleuropa. Vogelwarte 36(Sonderheft):1–221

Betts MG, Hadley AS, Rodenhouse N, Nocera JJ (2008) Social information trumps vegetation structure in breeding-site selection by a migrant songbird. Proc R Soc B 275:2257–2263

Bibby CJ, Green RE (1980) Foraging behaviour of migrant pied flycatchers, *Ficedula hypoleuca*, on temporary territories. J Anim Ecol 49:507–521

Bibby CJ, Green RE (1981) Autumn migration strategies of reed and sedge warblers. Ornis Scand 12:1–12

Biebach H (1990) Strategies of trans-Sahara migrants. In: Gwinner E (ed) Bird migration. Springer, Berlin

Biebach H, Friedrich W, Heine G (1986) Interaction of body mass, fat, foraging and stopover period in trans-Sahara migrating passerine birds. Oecologia 69:370–379

Bingman V (1980) Inland morning flight behavior of nocturnal passerine migrants in eastern New York. Auk 97:465–472

Bolshakov CV (1981) Rekonstruktsiya polnoy kartiny nochnogo prolyota i effektivnost' obnaruzheniya ego raznymi metodami (Reconstruction of the total picture of nocturnal passage and effectiveness of several methods of its estimation). Proc Zool Inst 104:95–123

Bolshakov CV, Shapoval AP, Zelenova NP (2001) Results of bird trapping and ringing by the Biological Station ''Rybachy'' on the Courish Spit: long-distance recoveries of birds ringed in 1956–1997. Part 1. Avian Ecol Behav (Suppl 1):1–126

Bolshakov CV, Shapoval AP, Zelenova NP (2002) Results of bird trapping and ringing by the Biological Station ''Rybachy'' on the Courish Spit: controls of birds ringed outside the courish spit in 1956–1997. Part 1. Avian Ecol Behav (Suppl 5):1–106

Bolshakov CV, Shapoval AP, Zelenova NP (2011) Results of bird trapping and ringing by the Biological Station "Rybachy" on the Courish Spit in 2010. Avian Ecol Behav 20:9–52

Bowlin MS, Cochran WW, Wikelski MC (2005) Biotelemetry of New World thrushes during migration: physiology, energetics and orientation in the wild. Integr Comp Biol 45:295–304

Brensing D (1989) Ökologische Untersuchungen der Tagesperiodik von Kleinvögeln. Ökol Vögel (Ecol Birds) 11:1–148

Buler JJ (2006) Understanding of habitat use by land birds during migration along the Mississippi Gulf coast using scale-dependent approach. Dissertation, University of Southern Mississippi

Buler JJ, Moore FR, Woltmann S (2007) A multi-scale examination of stopover habitat use by birds. Ecology 88:1789–1802

Bulyuk VN, Mukhin A, Fedorov VA, Tsvey A, Kishkinev D (2000) Juvenile dispersal in Reed Warblers *Acrocephalus scirpaceus* at night. Avian Ecol Behav 5:45–63

Cantos FJ, Tellería JL (1994) Stopover site fidelity of four migrant warblers in the Iberian Peninsula. J Avian Biol 25:131–134

Carpenter FL, Paton DC, Hixon MA (1983) Weight gain and adjustment of feeding territory size in migrant hummingbirds. Proc Natl Acad Sci USA 80:7259–7263

Carpenter FL, Hixon MA, Russel RW, Paton DC, Temeles EJ (1993a) Interference asymmetries among age-classes of rufous hummingbirds during migratory stopover. Behav Ecol Sociobiol 33:297–304

Carpenter FL, Hixon MA, Temeles EJ, Russel RW, Paton DC (1993b) Exploitative compensation by subordinate age-classes of migrant rufous hummingbirds. Behav Ecol Sociobiol 33:305–312

Catry P, Encarnação V, Araújo A, Fearon P, Fearon A, Armelin M, Delaloye P (2004) Are long-distance migrant passerines faithful to their stopover sites? J Avian Biol 35:170–181

Chernetsov N (1998) Habitat distribution during the post-breeding and post-fledging period in the reed warbler *Acrocephalus scirpaceus* and sedge warbler *A. schoenobaenus* depends on food abundance. Ornis Svecica 8:77–82

Chernetsov N (2005) Spatial behavior of medium and long-distance migrants at stopovers studied by radio tracking. Ann N Y Acad Sci 1046:242–252

Chernetsov N (2006) Habitat selection by nocturnal passerine migrants en route: mechanisms and results. J Ornithol 147:185–191

Chernetsov N, Bulyuk VN, Ktitorov P (2007) Migratory stopovers of passerines in an oasis at the crossroad of the African and Indian flyways. Ring Migr 23:243–251

Chernetsov N, Mukhin A (2006) Spatial behaviour of European robins *Erithacus rubecula* during migratory stopovers: a telemetry study. Wilson J Ornithol 118:364–373

Chernetsov N, Mukhin A, Ktitorov P (2004a) Contrasting spatial behaviour of two long-distance passerine migrants at spring stopovers. Avian Ecol Behav 12:53–61

Chernetsov NS, Skutina EA, Bulyuk VN, Tsvey AL (2004b) Optimal stopover decisions of migrating birds under variable stopover quality: model predictions and the field data. Zh Obschei Biologii 65:211–217

Chernetsov NS, Titov NV (2003) Pitanie i strategiya vesenney migratsii zaryanki, *Erithacus rubecula* (Aves, Turdidae), v yugo-vostochnoy Pribaltike (Foraging and spring migratory strategy of the Robin *Erithacus rubecula* (Aves, Turdidae) in the southeastern Baltic Sea region). Zool Zhurnal 82:1525–1529

Cochran WW, Bowlin MS, Wikelski M (2008) Wingbeat frequency and flap-pause ratio during natural migratory flight in thrushes. Integr Comp Biol 48:143–151

Cramp S (ed) (1992) The birds of the Western Palaearctic, vol 6. Cambridge University Press, Cambridge

Degen T, Jenni L (1990) Biotopnutzung von Kleinvögeln in einem Naturschutzgebiet und im umliegenden Kulturland während der Herbstzugzeit. Orn Beob 87:295–325

Deppe JL, Rotenberry JT (2008) Scale-dependent habitat use by fall migratory birds: vegetation structure, floristics, and geography. Ecol Monogr 78:461–487

Dierschke V (2003) Predation hazard during migratory stopover: are light or heavy birds under risk? J Avian Biol 24:24–29

Dolnik VR (1982) Problemy migratsiy ptits nad aridnymi i gornymi rayonami Sredney Azii (Problems of bird migration over arid and mountainous areas of Middle Asia). Ornithologia 17:13–17 (Moscow)

Dolnik VR (ed) (1985a) Energeticheskie resursy ptits, pereletayuschikh aridnye i gornye prostranstva Sredney Azii i Kazakhstana (Energy resources of birds migrating across arid and mountainous regions of Middle Asia and Kazakhstan). Zoological Institute, Leningrad

Dolnik VR (ed) (1985b) Nochnye migratsii ptits nad aridnymi i gornymi prostranstvami Sredney Azii i Kazakhstana (Nocturnal bird migration over arid and mountainous regions of Middle Asia and Kazakhstan). Zoological Institute, Leningrad

Dolnik VR (ed) (1987) Issledovanie migratsii ptits v aridnykh i gornykh rayonakh Sredney Azii i Kazakhstana (Study of bird migration in arid and mountainous regions of Middle Asia and Kazakhstan). Zoological Institute, Leningrad

Dolnik VR (1990) Bird migration across arid and mountainous regions of Middle Asia and Kasakhstan. In: Gwinner E (ed) Bird migration. Springer, Berlin

Dolnik VR, Bolshakov CV, Žalakevičius M (1981) Rekonstrukciya polnoy kartiny dnevnogo proleta i effektivnost' obnaruzheniya ee raznymi metodami (Reconstruction of the full picture of the diurnal passage and the efficiency of its detection by different methods). Proc Zool Inst 104:70–79

Dolnik VR, Yablonkevich ML (1985) Metody otlovov i prizhiznennogo obsledovaniya vorobyinych ptits v pustynnykh i gornykh rayonakh (Methods of capture and field examination of passerine birds in deserts and mountains). Proc Zool Inst 137:7–10

Dunn EH, Nol E (1980) Age-related migratory behavior of warblers. J Field Ornithol 51:254–269

Evans PR, Townsend DJ (1988) Site faithfulness of waders away from the breeding grounds: how individual migration patterns are established. In: Ouellet H (ed) Acta XIX Congr Int Ornithol, vol 1. University of Ottawa Press, Ottawa

Fox AD, Hilmarsson JO, Einarsson Ó, Walsh AJ, Boyd H, Kristiansen JN (2002) Staging site fidelity of Greenland white-fronted geese *Anser albifrons flavirostris* in Iceland. Bird Study 49:42–49

Fransson T, Barboutis C, Mellroth R, Akriotis T (2008) When and where to fuel before crossing the Sahara desert—extended stopover and migratory fuelling in first-year garden warblers *Sylvia borin*. J Avian Biol 39:133–138

Fransson T, Weber TP (1997) Migratory fuelling in blackcaps (*Sylvia atricapilla*) under perceived risk of predation. Behav Ecol Sociobiol 41:75–80

Freemark KE, Dunning JB, Hejl SJ, Probst JR (1995) A landscape ecology perspective for research, conservation, and management. In: Martin TE, Finch DM (eds) Ecology and management of Neotropical migratory birds: a synthesis and review of critical issues. Oxford University Press, NY

Gauthreaux SA Jr (1978) Importance of daytime flights of nocturnal migrants: redetermined migration following displacement. In: Schmidt-Koenig K, Keeton WT (eds) Animal migration, navigation, and homing. Springer, Berlin

Gavrilov EI (1998) Individualnye sroki proleta indiyskogo i chernogrudogo vorobyev v predgoryakh Zapadnogo Tian-Shanya (Individual timing of passage of Indian and Spanish sparrows in the foothills of Western Tien Shan). In: Kurochkin EN (ed) Sovremennaya ornitologiya (Current ornithology) 1998. Nauka, Moscow

Grünberger S, Leisler B (1990) Angeborene und erfahrungsbedingte Komponenten der Habitatwahl der Tannenmeise (*Parus ater*). J Ornithol 131:460–464

Grünberger S, Leisler B (1993) Die Ausbildung von Habitatpräferenzen bei der Tannenmeise (*Parus ater*): genetische Prädisposition und Einfluß der Jugenderfahrung. J Ornithol 134:355–358

Hansen L (1954) Birds killed at lights in Denmark 1886–1939. Vidensk Medd den naturhist Foren i København 116:269–368

Hansson M, Pettersson J (1989) Competition and fat deposition in Goldcrests (*Regulus regulus*) at a migration stopover site. Vogelwarte 35:21–31

Hedenström A, Alerstam T (1997) Optimum fuel loads in migratory birds: distinguishing between time and energy minimization. J Theor Biol 189:227–234

Heise CD, Moore FR (2003) Age-related differences in foraging efficiency, molt, and fat deposition of gray catbirds prior to autumn migration. Condor 105:496–504

Herremans M (1990a) Can night migrants use interspecific song recognition to assess habitat? Gerfaut 80:141–148

Herremans M (1990b) Body-moult and migration overlap in reed warblers (*Acrocephalus scirpaceus*) trapped during nocturnal migration. Gerfaut 80:149–158

Hinsley SA, Bellamy PE, Newton I, Sparks TH (1995) Habitat and landscape factors influencing the presence of individual breeding bird species in woodland fragments. J Avian Biol 26:94–104

Houston AI (1998) Models of optimal avian migration: state, time and predation. J Avian Biol 29:395–404

Hutto RL (1985a) Seasonal changes in the habitat distribution of transient insectivorous birds in south-eastern Arizona: competition mediated? Auk 102:120–132

Hutto RL (1985b) Habitat selection by nonbreeding, migratory land birds. In: Cody ML (ed) Habitat selection in birds. Academic Press, NY

Jenni L (1996) Habitatwahl nachtziehender Kleinvögel bei Bodennebel. J Ornithol 137:425–434

Karandina SN, Erpert SD (1972) Klimaticheskoe ispytanie drevesnykh porod v Prikaspiyskoy polupustyne (Climatic testing of tree species in the Caspian semi-desert). Nauka, Moscow

Kelly JF, Delay LS, Finch DM (2002) Density-dependent mass gain by Wilson's warblers during stopover. Auk 119:210–213

Kerlinger P, Moore FR (1989) Atmospheric structure and avian migration. In: Johnston RF (ed) Current ornithology, vol 6. Plenum Press, NY

Kodric-Brown A, Brown JH (1978) Influence of economics, interspecific competition, and sexual dimorphism on territoriality in migrant rufous hummingbirds. Ecology 49:285–296

Ktitorov P, Bairlein F, Dubinin M (2008) The importance of landscape context for songbirds on migration: body mass gain is related to habitat cover. Landsc Ecol 23:169–179

Ktitorov P, Tsvey A, Mukhin A (2010) The good and the bad stopover: behaviours of migrant reed warblers at two contrasting sites. Behav Ecol Sociobiol 65:1135–1143

Lank D, Ydenberg R (2003) Death and danger at migratory stopovers: problems with 'predation risk'. J Avian Biol 34:225–228

Leisler B, Ley H-W, Winkler H (1989) Habitat, behaviour and morphology of *Acrocephalus* warblers: an integrated analysis. Ornis Scand 20:181–186

Leisler B, Schulze-Hagen K (2011) The reed warblers. KNNV Publishing, Zeist

Ley H-W (1988) Verhaltensontogenese der Habitatwahl beim Teichrohrsänger (*Acrocephalus scirpaceus*). J Ornithol 129:287–297

Loria DE, Moore FR (1990) Energy demands of migration on red-eyed vireos, *Vireo olivaceus*. Behav Ecol 1:24–35

Mädlow W (1997) Durchzug und Rastverhalten des Rotkehlchens (*Erithacus rubecula*) im Herbst 1995 auf der Greifswalder Oie: situation während eines Masseneinzugs. Seevögel 18:75–81

Malchesvky AS, Pukinsky YB (1983) Ptitsy Leningradskoy oblasti i sopredelnykh territoriy (The birds of Leningrad region and adjacent territories). Leningrad State University Press, Leningrad

Martin TE (1980) Diversity and abundance of spring migratory birds using habitat islands on the Great Plains. Condor 82:430–439

Martin TE, Karr JR (1986) Patch utilization by migrating birds: resource oriented? Ornis Scand 17:165–174

Merom K, Yom-Tov Y, McClery R (2000) Philopatry to stopover site and body condition of transient reed warblers during autumn migration through Israel. Condor 102:441–444

Mettke-Hofmann C, Gwinner E (2004) Differential assessment of environmental information in a migratory and a nonmigratory passerine. Anim Behav 68:1079–1086

Mills AM, Thurber BJ, Mackenzie SA, Taylor PD (2011) Passerines use nocturnal flight for landscape-scale movements during migratory stopover. Condor 113:597–607

Mönkkönen M, Härdling R, Forsman JT, Tuomi J (1999) Evolution of heterospecific attraction: using other species as cues in habitat selection. Evol Ecol 13:91–104

Moore FR (1987) Sunset and the orientation behaviour of migrating birds. Biol Rev 62:65–86

Moore FR (1990) Evidence for redetermination of migratory direction following wind displacement. Auk 107:425–428

Moore FR, Aborn DA (2000) Mechanisms of en route habitat selection: how do migrants make habitat decisions during stopover? Stud Avian Biol 20:34–42
Moore FR, Gauthreaux SA Jr, Kerlinger P, Simons TR (1993) Stopover habitat: management implications and guidelines. In: Finch DM, Stangel PW (eds) Status and management of Neotropical migratory birds. General Technical Report RM-229, Rocky Mountain Research Station, Forest Service, U.S.D.A., Fort Collins
Moore FR, Gauthreaux SA Jr, Kerlinger P, Simons TR (1995) Habitat requirements during migration: important link in conservation. In: Martin TE, Finch DM (eds) Ecology and management of Neotropical migratory birds: a synthesis and review of critical issues. Oxford University Press, NY
Moore FR, Kerlinger P, Simons TR (1990) Stopover on a Gulf coast barrier island by spring trans-Gulf migrants. Wilson Bull 102:487–500
Moore FR, Mabey S, Woodrey M (2003) Priority access to food in migratory birds: age, sex and motivational asymmetries. In: Berthold P, Gwinner E, Sonnenschein E (eds) Avian migration. Springer, Berlin
Moore FR, Wang Y (1991) Evidence of food-based competition among passerine migrants during stopover. Behav Ecol Sociobiol 28:85–90
Moore FR, Woodrey MS, Buler JJ, Woltmann S, Simons TR (2005) Understanding the stopover of migratory birds: a scale-dependent approach. In: Ralph CJ, Rich TD (eds) Bird conservation implementation and integration in the Americas: Proceedings of the 3rd International Partners in Flight conference 2002, USDA Forest Service General Technical Report PSW-191, Albany
Mouritsen H (2003) Spatiotemporal orientation strategies of long-distance migrants. In: Berthold P, Gwinner E, Sonnenschein E (eds) Avian migration. Springer, Berlin
Mukhin A (2004) Night movements of young reed warblers (*Acrocephalus scirpaceus*) in summer: is it postfledging dispersal? Auk 121:203–209
Mukhin AL, Chernetsov NS, Kishkinev DA (2005) Pesnya trostnikovoy kamyshevski, *Acrocephalus scirpaceus* (Aves, Sylviidae), kak akusticheskiy marker vodno-bolotnogo biotopa vo vremya migratsii (Reed warbler, *Acrocephalus scirpaceus* (Aves, Sylviidae), song as an acoustic marker of wetland biotope during migration). Zool Zhurnal 85:995–1002
Mukhin A, Chernetsov N, Kishkinev D (2008) Acoustic information as a distant cue for habitat recognition by nocturnally migrating passerines during landfall. Behav Ecol 19:716–723
Ormerod SJ (1990) Time of passage, habitat use and mass change of *Acrocephalus* warblers in a South Wales reedswamp. Ring Migr 11:1–11
Ottich I, Dierschke V (2003) Exploration of resources modulates stopover behaviour of passerine migrants. J Ornithol 144:307–316
Panov IN (2008) Sluchay vernosti puti migratsii u varakushki (A case of fidelity to the migratory route in the bluethroat). Ornithologia 35:146–147 (Moscow)
Panov IN (2012) Migratsionnye strategii vorobyinykh ptits v severnoy tayge Vostochnoy Fennoskandii (Migratory strategies of passerines in the northern taiga of Eastern Fennoscandia). Dissertation, Institute of Ecology and Evolution RAS
Panov IN, Chernetsov NS (2010a) Migratsionnaya strategiya varakushki (*Luscinia svecica*) v Vostochnoy Fennoskandii. Soobschenie 1: Osnovnye parametry migratsionnykh ostanovok (Migratory strategy of bluethroats, *Luscinia svecica,* in Eastern Fennoscandia. Part 1: Main stopover parameters). Proc Zool Inst 314:93–104
Panov IN, Chernetsov NS (2010b) Migratsionnaya strategiya varakushki (*Luscinia svecica*) v Vostochnoy Fennoskandii. Soobschenie 2: Reaktsiya na akusticheskie markery i vybor biotopa na migratsionnoy ostanovke (Migratory strategy of bluethroats, *Luscinia svecica,* in Eastern Fennoscandia. Part 2: Response to acoustic markers and habitat selection at stopover). Proc Zool Inst 314:173–183
Parnell JF (1969) Habitat relations of the parulidae during spring migration. Auk 86:505–521
Payevsky VA (1985) Demografiya ptits (Avian demography). Nauka, Leningrad
Payevsky VA (1998) Age structure of passerine migrants at the Eastern Baltic coast: the analysis of the 'coastal effect'. Ornis Svecica 8:171–178

Payevsky VA (1999) Adaptive significance of seasonal migrations: are annual migrations dangerous for birds? Russ J Zool 3:78–84
Payevsky VA (2009) Songbird demography. Pensoft, Sofia, Moscow
Pfister C, Kasprzyk MJ, Harrington BA (1998) Body fat levels and annual return in migrating semipalmated sandpipers. Auk 115:904–915
Pienkowski MW (1976) Recurrence of waders on autumn migration at sites in Morocco. Vogelwarte 28:293–297
Ralph CJ (1978) Disorientation and possible fate of young passerine coastal migrants. Bird-Banding 49:237–247
Ralph CJ (1981) Age ratios and their possible use in determining autumn routes of passerine migrants. Wilson Bull 93:164–188
Rappole JH, Warner DW (1976) Relationships between behavior, physiology and weather in avian transients at a migration stopover site. Oecologia 26:193–212
Salewski V, Almasi B, Heuman A, Thoma M, Schlageter A (2007) Agonistic behaviour of Palaearctic passerine migrants at a stopover site suggests interference competition. Ostrich 78:349–355
Salewski V, Schmaljohann H, Liechti F (2010) Spring passerine migrants stopping over in the Sahara are not fall-outs. J Ornithol 151:371–378
Sapir N, Tsurim I, Gal B, Abramsky Z (2004) The effect of water availability on fuel deposition of two staging *Sylvia* warblers. J Avian Biol 35:25–32
Schaub M, Jenni L (2000) Fuel deposition of three passerine bird species along migration route. Oecologia 122:306–317
Schaub M, Schwilch R, Jenni L (1999) Does tape-luring of migrating Eurasian reed warblers increase number of recruits or capture probability? Auk 116:1047–1053
Schlaepfer MA, Runge MC, Sherman PW (2002) Ecological and evolutionary traps. Trends Ecol Evol 17:474–480
Schmaljohann H, Dierschke V (2005) Optimal bird migration and predation risk: a field experiment with northern wheatears *Oenanthe oenanthe*. J Anim Ecol 74:131–138
Shochat E, Abramsky Z, Pinshow B, Whitehouse MAE (2002) Density-dependent habitat selection in migratory passerines during stopover: what causes the deviation from IFD? Evol Ecol 16:469–488
Taylor PD, Mackenzie SA, Thurber BG, Calvert AM, Mills AM, McGuire LP, Guglielmo CG (2011) Landscape movements of migratory birds and bats reveal and expanded scale of stopover. PLoS ONE 6:e27054
Thomson RL, Forsman JT, Mönkkönen MR (2003) Positive interactions between migrant and resident birds: testing the heterospecific attraction hypothesis. Oecologia 134:431–438
Titov N (1999a) Individual home ranges of Robins *Erithacus rubecula* at stopovers during autumn migration. Vogelwelt 120:237–242
Titov N (1999b) Fat level and temporal pattern of diurnal movements of Robins (*Erithacus rubecula*) at an autumn stopover site. Avian Ecol Behav 2:89–99
Titov N (1999c) Home ranges in two passerine nocturnal migrants at a stopover site in autumn. Avian Ecol Behav 3:69–78
Titov N (2000) Interaction between foraging strategy and autumn migratory strategy in the Robin *Erithacus rubecula*. Avian Ecol Behav 5:35–44
Tsvey AL (2008) Strategii migratsii zaryanki (*Erithacus rubecula*) v vostochnoy Pribaltike (Migratory strategies of the European robin (*Erithacus rubecula*) in the eastern Baltic). Dissertation, Zoological Institute RAS
Veiga JP (1986) Settlement and fat accumulation by migrant pied flycatchers in Spain. Ring Migr 7:85–98
Wang Y, Finch DM, Moore FR, Kelly JF (1998) Stopover ecology and habitat use of migratory Wilson's warbler. Auk 115:839–842
Wang Y, Moore FR (2005) Long-distance bird migrants adjust their foraging behavior in relation to energy stores. Acta Zool Sin 51:12–23

Weber TP, Houston AI (1997a) Flight costs, flight range and the stopover ecology of migrating birds. J Anim Ecol 66:297–306

Weber TP, Houston AI (1997b) A general model for time-minimising avian migration. J Theor Biol 185:447–458

Wiedner DS, Kerlinger P, Sibley DA, Holt P, Hough J, Crossley R (1992) Visible morning flights of Neotropical landbird migrants at Cape May, New Jersey. Auk 109:500–510

Winkler H, Leisler B (1985) Morphological aspects of habitat selection in birds. In: Cody ML (ed) Habitat selection in birds. Academic Press, San Diego

Woodrey M (2000) Age-dependent aspects of stopover biology of passerine migrants. Stud Avian Biol 20:43–52

Yaukey PH (2010) Concentrated migratory morning flight at Lake Pontchartrain, Louisiana, USA. Wilson J Ornithol 122:738–743

Yosef R, Chernetsov N (2004) Stopover ecology of migratory sedge warblers (*Acrocephalus schoenobaenus*) at Eilat, Israel. Ostrich 75:52–56

Yosef R, Chernetsov N (2005) Longer is fatter: body mass changes of migrant reed warblers (*Acrocephalus scirpaceus*) staging at Eilat, Israel. Ostrich 76:142–147

Chapter 6
Spatial Behaviour at Stopovers

Abstract This chapter deals with the range, direction and pattern of diurnal movements of nocturnal passerine migrants at stopovers. Contrary to the popular assumption, these movements are shown not to be directed generally towards the goal of migration, but to be purely habitat-related. The results of the original studies of territoriality vs. broad movements at stopovers are reported together with the critical re-analysis of literature data. I discuss the benefits and pitfalls of visual observations of marked individuals, capture-recapture studies and radio tracking. I also analyse an ambiguous impact of current fuel stores on spatial behaviour of songbirds at stopover. The main factor that governs spatial use of stopover migrants is the spatial distribution of food, clumped versus relatively uniform. If the food is uniformly distributed in space and predictable in time, migrants occupy and sometimes defend small home ranges. The species whose preferred food varies broadly in space and time, make broad movements at stopovers more often than not.

6.1 Introductory Remarks

Habitat selection and use by migrants is closely related to their spatial behaviour. It should be specially emphasized that even though these two aspects are closely related, they are not identical. Some songbird migrants occupy temporary territories at stopovers (Rappole and Warner 1976; Kodric-Brown and Brown 1978; Bibby and Green 1980, 1981; Carpenter et al. 1983, 1993a, b) or remain within very restricted areas, but do not defend them (Titov 1999a, b; Lajda 2001; Chernetsov and Titov 2001a; Chernetsov 2002, 2005; Chernetsov and Mukhin 2006). Other species move broadly during stopovers (Aborn and Moore 1997; Chernetsov et al. 2004). The same species may occupy small restricted home ranges in some situations [Eurasian reed warblers on the Courish Spit (Chernetsov

N. Chernetsov, *Passerine Migration*, DOI: 10.1007/978-3-642-29020-6_6,

and Titov 2001a) and pied flycatchers in Portugal (Bibby and Green 1980)] and move broadly in others [Eurasian reed warblers in Portugal and France (Bibby and Green 1981) and pied flycatchers on the Courish Spit (Chernetsov et al. 2004)]. Habitat use may be very similar whereas spatial behaviour may differ. In the same habitat spatial behaviour may be density-related.

Recent radio-tracking data shows that songbird migrants that normally remain within limited areas during stopovers (within tens or a few hundreds of metres) may make longer landscape-scale movements (5–30 km) at night (Mills et al. 2011; Taylor et al. 2011). It is probably a matter of taste whether to consider that such movements occur within stopover, or to classify them as very short movements between different stopover sites. I suggest the former variant.

6.2 Range and Direction of Daytime Movements of Nocturnal Migrants

One of the key questions is the scale of diurnal movements of nocturnal migrants. Is it hundreds of metres, kilometres, tens of kilometres? A correct estimate of the scale of these movements can allow a reasonable guess of their function. If the movements occur within hundreds of metres or several kilometres, they are most probably aimed at foraging optimisation. If stopover migrants routinely cover tens of kilometres during their daytime movements, they may be aiming to move towards the migratory target. Therefore, the question of the scale is linked to the question whether or not nocturnal migrants move predominantly in the migratory direction during their daytime stopovers. It is often claimed that they do (Hantge and Schmidt-Koenig 1958; Schmidt-Koenig 1980; Schlenker 1980; Shumakov 1985). Even though observations that support this view are often anecdotal, this viewpoint seems self-evident and not needing any hard proof. It is shared by many bird migration researchers even today. For example, the new method of studying orientation of nocturnal migrants in the so-called Busse cage (Busse 1995, 2000) is based on the assumption that nocturnally migrating songbirds are motivated to move (and apparently do move) in the migratory direction during the daylight hours. Otherwise it is impossible to explain why orientation tests with nocturnal migrants are performed in the daytime.

Captures at trapping stations that have a necessarily restricted area have a limited value for solving this problem. This is especially true for the captures into stationary funnels traps of Helgoland or Rybachy type: stopover migrants that have moved after capture for only 50–100 m leave the recruitment area, so that their recapture probability drops nearly to nil. Mist-netting sites, especially when mist-nets are spread across a relatively large area, are better suited for studying spatial behaviour of migrants (Chernetsov and Titov 2000). However, even in such cases the distance between the furthest nets for practical reasons usually does not exceed several hundred metres.

To estimate the possible maximum distance of movements, it is necessary to analyse data from two or several trapping stations located near each other. I analysed 123 recaptures of sedge warblers at several sites in western Estonia, where in the 1980s more than a dozen trapping stations were operating on the Baltic coast 2–20 km from one another (Chernetsov 1998b). Until early August, recaptures occurred both south and north of the capture site, with 35% of birds moving north (n = 85). Only one bird produced a long-distance recovery during this period (on August 6, 1990 in the former Czechoslovakia). These data suggest that most birds were performing dispersal movements without obvious directionality, and only a fraction had already started southward migration. Since mid August, recaptures south of the trapping site comprised 84% (n = 38), i.e. most movements occurred towards the south. It seems that at the end of the postfledging period and at the beginning of autumn passage of sedge warblers, they may indeed move towards the goal of migration by small hops. However, it remained open whether sedge warblers performed very short nocturnal flights, or they moved south in the daytime.

I also analysed recaptures of six species of nocturnal migrants (European robin, willow warbler, chiffchaff, blackcap, garden warbler, pied flycatcher) ringed at one of the two field stations on the Courish Spit (Rybachy and Fringilla) and recaptured at the other site. Unlike the recent paper (Chernetsov 2011), here I include into analysis birds that were recaptured within 15 days, and not only within-day recaptures. Individuals that looked ill or injured were excluded, as their behaviour might not have represented the patterns followed by healthy stopover migrants. In some species, periods of postfledging movements and autumn migration may overlap, when late hatched local individuals are still performing postfledging movements in the broad vicinity of their natal sites, and early hatched individuals from further north are already migrating through the study area. Therefore we also excluded from the analysis a few birds in heavy moult.

Besides calculating the proportion of birds switching between the two capture sites, we also tested whether the numbers of individuals moving in the migratory and in the reverse direction differed significantly from 1:1 ratio. This was done to check whether these movements occurred significantly more often in the migratory direction.

Trapping figures of study species in 1993–2006, when both sites were operating in parallel, are presented in Table 6.1. The total number of European robins that switched between the sites was 32, 11 birds did it in spring and 21 in autumn (Table 6.2). In spring, five robins moved in the migratory and six in the reverse direction, but the percentage of birds ringed at the respective sites was higher for the migratory direction. The difference was marginally insignificant (Table 6.3). In autumn, conversely, a significantly greater proportion of European robins moved in the reverse direction than in the migratory one (Table 6.3).

Only three European robins covered the distance between capture sites within a single day, i.e. definitely during daytime movements. All of them did it in autumn in the reverse direction. Some other birds could also have taken off but aborted flight, e.g. due to adverse wind conditions. This was the case with the European

Table 6.1 Trapping figures of long-distance migrants at the field stations Rybachy and Fringilla on the Courish Spit in 1993–2006

	European robin	Willow warbler	Chiffchaff	Blackcap	Garden warbler	Pied flycatcher
Rybachy						
Spring	37,325	2,132	2,088	1,770	942	1,783
Autumn	47,652	7,591	1,262	6,818	4,308	1,378
Fringilla						
Spring	9,085	3,566	1,013	705	163	944
Autumn	17,415	17,556	870	1,089	1,597	1,249

robin that was radio-tagged in Fringilla in the morning of 15 April 2005, spent the day in the vicinity of the captured site, took off after dark and was recaptured in the morning of 16 April in Rybachy, 10.7 km northeast of the departure site.

The proportions of willow warblers that moved in the migratory and in the reverse direction were not significantly different in either season (Table 6.3). Five willow warblers covered the distance between the capture sites during a single day, all in spring in the migratory direction (Table 6.2). A significantly greater proportion of chiffchaffs moved in the migratory than in the reverse direction in spring, in autumn the difference was not significant (Table 6.3). Two individuals covered the distance between the sites within a single day; both did in spring, one in each direction (Table 6.2).

Of 8,588 blackcaps captured in Rybachy and 1,794 captured in Fringilla, one bird moved between the sites in spring and one in autumn (Table 6.2). It should however be noted that the autumn bird took 8 days to cover the distance between the sites, and it was at the final stage of juvenile moult of body feathers. It cannot therefore be ruled out that this individual was at the stage of dispersal rather than migration (however, the last stage of body moult does not rule out autumn migration, either). The spring bird was a yearling female that had covered 10.7 km in the reverse direction in five hours. With so few birds, no direction was significantly preferred in either season (Table 6.3).

Of 5,250 garden warblers captured in Rybachy and 1,760 captured in Fringilla in the same period, none was recaptured between these two sites. Of 3,161 pied flycatchers ringed in Rybachy and 2,193 in Fringilla, only one individual moved between the sites, from Fringilla to Rybachy in autumn in five hours.

All these species are purely nocturnal migrants (Bolshakov 1977). In 1993–2006, a total of 102,546 goldcrests were ringed in Fringilla and 43,119 in Rybachy. Of them, 116 individuals, or 0.080%, were recorded switching sites. The goldcrest is not a purely nocturnal migrant; this species has a diurnal phase of migratory movements (Bolshakov 1977). Fourteen goldcrests moved between the trapping stations in spring and 102 in autumn. In spring five birds moved in the migratory direction and nine in the reserve direction (Yates-corrected $\chi^2 = 0.15$, $p = 0.70$), and in autumn the figures were 83 and 19, respectively (Yates-corrected $\chi^2 = 20.9$,

Table 6.2 Numbers of long-distance migrants that moved between Rybachy and Fringilla on the Courish Spit (10.7 km) in the migratory and in the reverse direction

	European robin	Willow warbler	Chiffchaff	Blackcap	Garden warbler
Spring	**11**	**9**	**6**	**1**	
Migratory direction	5	7	5		
0 nights		5	1		
1 night	3		3		
>1 nights	2	2	1		
Reverse direction	6	2	1	1	
0 nights			1	1	
1 night	4	1			
>1 nights	2	1			
Autumn	**21**	**6**	**3**	**1**	**1**
Migratory direction	8	4	1		
0 nights					
1 night	2	3			
>1 nights	6	1	1		
Reverse direction	13	2	2	1	1
0 nights	3				1
1 night	5		1		
>1 nights	5	2	1	1	

Table 6.3 Proportions of birds that moved in the migratory and the reverse direction between Rybachy and Fringilla and the difference between the proportions (Yates-corrected χ^2)

Вид	European robin	Willow warbler	Chiffchaff	Blackcap	Garden warbler	European robin
Spring	3.18; $p = 0.075$	0.36; $p = 0.55$	4.90; $p = 0.027$	0.23; $p = 0.63$		
Migratory	0.055%	0.196%	0.494%	0	0	0
Reverse	0.016%	0.094%	0.048%	0.056%	0	0
Autumn	11.50; $p = 0.0007$	2.26; $p = 0.13$	0.11; $p = 0.75$	1.11; $p = 0.29$		0.00; $p = 0.96$
Migratory	0.017%	0.053%	0.079%	0	0	0
Reverse	0.075%	0.011%	0.230%	0.092%	0	0.080%

$p < 0.001$). In the species that performs migratory flights both at night and by day, movements were significantly directed towards the winter quarters in autumn.

Percentage of nocturnal passerines migrants that were recaptured during the migratory period between two sites located 10.7 km from each other varied between 0 (garden warbler in both directions and seasons) and 0.494% (chiffchaff in the migratory direction in spring). Even though percentages varied between different species, all of them were remarkably low.

It should be emphasized that these figures are probably overestimates: at least some individuals that moved between the trapping sites did not do so by daytime movements. At least one European robin ringed in Fringilla and recaptured in

Rybachy on the next day covered the distance between the sites by the nocturnal flight that probably was initiated and later aborted due to unfavourable wind conditions (Schmaljohann et al. 2011) or was a landscape-scale nocturnal movement aimed at locating better stopover habitats (Mills et al. 2011). Some of the remaining birds whose movement included at least one night could have done the same. Therefore, the number of birds that actually moved between the trapping stations in the daytime could probably have been even lower than inferred from our data. In fact, only those individuals that moved between the sites within a single day (five willow warblers, three European robins, two chiffchaffs, one blackcap and one pied flycatcher) must have covered 10.7 km in the daytime.

Radio-tagging data suggest that the scale of movements at stopovers is measured by hundreds of metres, at most several kilometres (Aborn and Moore 1997; Chernetsov et al. 2004; Chernetsov and Mukhin 2006; Fransson et al. 2008; Mills et al. 2011). Our trapping data from the Courish Spit clearly do not support the idea that diurnal movements of nocturnal migrants are generally directed towards the migratory target and bring them considerably closer to it. Our data support the results of analysis of movements at a single stopover site in SW Germany by Bastian (1992) who found that these movements occurred randomly in different directions. Tsvey (2008) did not find any preference for a specific direction of movements in radio-tagged European robins on the Courish Spit, either in spring or in autumn.

Taken together, all these data strongly suggest that in most cases the aim of daytime movements of nocturnal migrants at stopovers is foraging optimisation and search for optimal habitats. Nocturnal migratory flight is a totally different form of behaviour. This is also supported by the studies of diel patterns of activity of caged migrants: diurnal activity is always followed by a quiescence period before the nocturnal peak of activity starts (Gwinner 1996; Ramenofsky et al. 2003; Agatsuma and Ramenofsky 2006). This is further supported by the data on orientation of European robins that did not show preference of their locomotor activity in the daytime in the seasonally appropriate direction, and their orientation was not influenced by magnetic field manipulations (Wiltschko and Höck 1972).

6.3 Restricted Home Ranges Versus Broad Movements: Visual Observations and Recapture Analysis

6.3.1 Visual Observations and Their Limitations

The initial method of studying spatial behaviour of passerines at stopovers was visual observation of individually marked migrants. This method has shown that pied flycatchers in Portugal defend temporary territories during autumn passage (Bibby and Green 1980), and two *Acrocephalus* species, Eurasian reed warbler and sedge warbler, behave differently (Bibby and Green 1981). Sedge warblers mainly feed on plum aphids, actively move at stopover and often feed close to conspecifics; whereas Eurasian reed warblers that take many dipterans, occupy non-overlapping home ranges that are at least sometimes defended, i.e. are territories. Unfortunately, the authors give no quantitative data supporting this claim. Furthermore, our studies of diet of *Acrocephalus* warblers during autumn passage in the Baltic area (Chernetsov and Manukyan 1999a, b, 2000) showed a much greater overlap between diets of these two species than found by Bibby and Green (1981) in Portugal and France. Our data suggest that when plum aphids are abundant (which does not happen each year), they are equally readily taken by both species. As the diet of both species seems to be more catholic than usually assumed, their spatial behaviour may also be actually less consistently different than suggested.

The study of spatial behaviour of songbirds at stopovers by direct visual observation is methodologically very difficult. The problem is not only that it is not straightforward to observe repeatedly individually marked songbirds during the brief migratory stopover. As many passerine nocturnal migrants are woodland birds (and the recently mentioned *Acrocephalus* warblers inhabit reedbeds), it is difficult enough. It is further complicated by the fact that at stopovers many passerines often used habitats at early succession stages (Bairlein 1981) in which mosaicity is very high. It is virtually impossible to ensure equal detection probability of a bird in different patches of a fine-grain mosaic habitat by visual observations.

6.3.2 Analysis of Recaptures and Its Limitations

In order to tackle this methodological problem, we have developed a method to estimate spatial behaviour on the basis of analysis of recaptures in standardised mist-netting projects (Titov 1999a; Chernetsov and Titov 2001b). The idea of the method is to analyse frequency distribution of distances between capture sites of the same individual. If mist-net locations are known and held constant and the net number is recorded during capture, the capture site is known to the nearest several metres. It makes it possible to calculate distances between all the pairs of nets operated at the trapping station. The frequency distribution of distances between capture and recapture sites can be compared with the neutral model that assumes random capture at any site within the trapping state, i.e. in any net. The neutral model is calculated by simulating captures in virtual 'nets'. This simulation may take the habitat use of the species in question into account by making the probability of drawing a certain net equal to the proportion of real captures in this net. It makes the neutral model much more realistic.

Frequency distribution of distances between real captures is compared with the simulated distribution which is constructed under the assumption that birds move randomly across the study site and, habitat use taken in account, are randomly captured in any net. This is what the neutral model approach is, when impact of some factor is studied by comparing the real situation with the model assuming

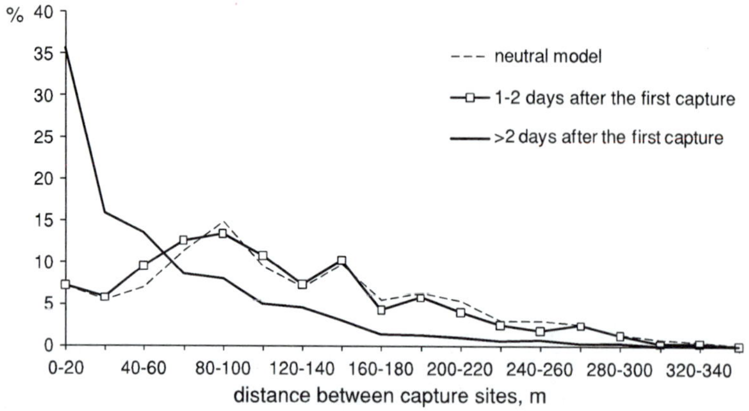

Fig. 6.1 The frequency distribution of distances between the capture locations of European robins in autumn on the Courish Spit. Recaptures within 1–2 days after the initial capture; recaptures >2 days after the initial capture and the neutral model of random movements across the trapping site are shown

that this factor is not acting (Begon et al. 2006). If the two distributions, the real and the simulated one, are not significantly different, there is no reason to reject the null hypothesis that home ranges of the birds studied are not smaller than the whole study area. Usually it means that home ranges are too large to be defended territories and are jointly used by several (often many) individuals.

If in reality the birds are recaptured significantly closer to the site of their previous capture than predicted by the neutral model, it means that they remain within home ranges that are significantly smaller than the whole study area. This method does not make it possible to find out whether these home ranges are defended territories: it is only possible through direct behavioural observations. A hypothetic situation when recaptures occur significantly farther from the previous capture site than predicted by the random model would mean that birds avoid the site of their capture. This situation has not yet been reported (Titov 1999a, c; Chernetsov and Titov 2001a; Lajda 2001; Chernetsov 2002), but theoretically it cannot be ruled out.

Recapture data on European robins in Rybachy in 1994–1996 shows that during the first two days after the initial capture no significant difference was recorded between the distribution of real captures and the neutral model (Wilcoxon matched pairs test, $z = 0.24$, $p = 0.81$; Fig. 6.1). However, captures on the third and subsequent days occurred significantly closer to the site of the previous capture than predicted by the model ($z = 2.59$, $p = 0.009$, Fig. 6.1). It means that in the first two days European robins moved broadly across the study site (Cape Rossitten), and since the third day of stopover the birds remained within home ranges whose area was significantly smaller than the trapping station.

Similar patterns have been shown by the winter wren *Troglodytes troglodytes* and dunnock *Prunella modularis*. The median distances between capture locations on

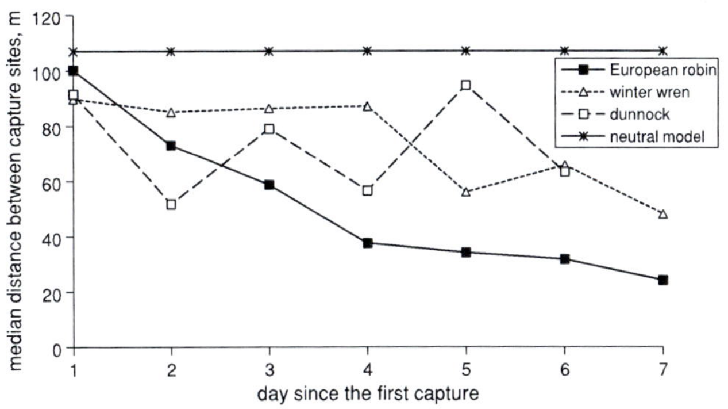

Fig. 6.2 The median distance between the capture sites in relation to the day of stopover in European robins, winter wrens and dunnocks during autumn migratory stopovers on the Courish Spit

different days since arrival in European robins, winter wrens and dunnocks are shown in Fig. 6.2. In European robins and winter wrens the scale of movements declined during their stay. In all three species the observed pattern clearly differed from the predictions of the neutral model (Wilcoxon matched pairs test, $z = 2.37$, $p = 0.018$ for both European robin and winter wren; $z = 2.20$, $p = 0.028$ for the dunnock).

A different pattern was recorded in the Eurasian reed and sedge warblers (Chernetsov and Titov 2001a). Both Eurasian reed warblers that had completed moult (probably transient migrants from more northern populations; Chernetsov 1998c) and birds in moult (that included local individuals) moved across the study site randomly (within their preferred habitats): $z = 1.12$, $p = 0.26$ (Fig. 6.3) and $z = 0.97$, $p = 0.33$ (Fig. 6.4). The same pattern was observed in sedge warblers during postfledging movements and autumn migratory stopovers ($z = 0.77$, $p = 0.44$, Fig. 6.5). Unfortunately, in this species body moult is so reduced (except for south European populations, Zehtindjiev 1989) that its pattern does not make it possible to distinguish between local individuals performing postfledging movements and transient migrants at stopover (Fedorov 1990; Redfern and Alker 1996).

It should be emphasized that broad movements of *Acrocephalus* warblers were not only typical of the initial period of their stay at the study site and were not a result of longer stopovers than in European robins. The pattern of movements, as shown by frequency distribution of distances between capture locations, did not differ in reed warblers that stayed for 1–3 days and for more than three days: Friedmann ANOVA, $F_{2,9} = 4.67$, $p > 0.05$ (Fig. 6.6).

Similar results were obtained for the blackcap in Rybachy (Chernetsov 2002). Their movements were not significantly different from random ones within the area covered by mist-nets either in moulting individuals (Wilcoxon matched pairs test, $z = 0.35$, $p = 0.73$) or in birds that had completed moult and were probably on migration ($z = 0.03$, $p = 0.99$). All the three warbler species moved broadly across our study site, and birds that were simultaneously present on

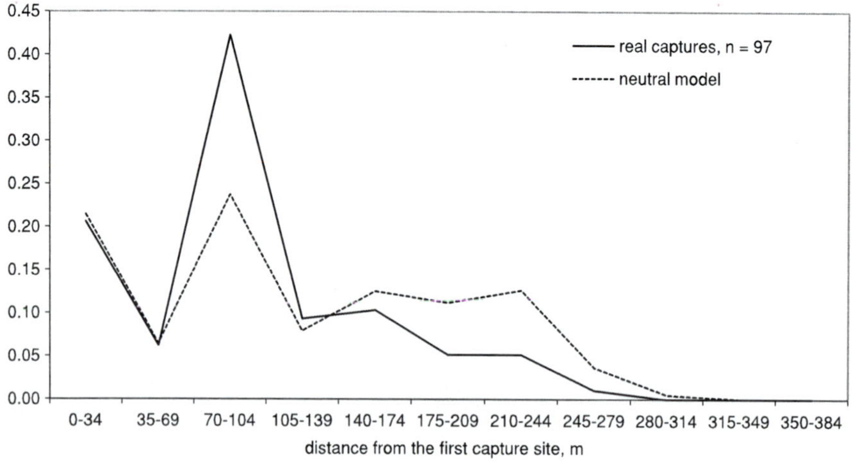

Fig. 6.3 The frequency distribution of distances between the capture sites of Eurasian reed warblers with completed moult (i.e. on migration) in autumn on the Courish Spit and the neutral model. The difference is not significant (Wilcoxon matched pairs test, $z = 1.12$, $p = 0.26$)

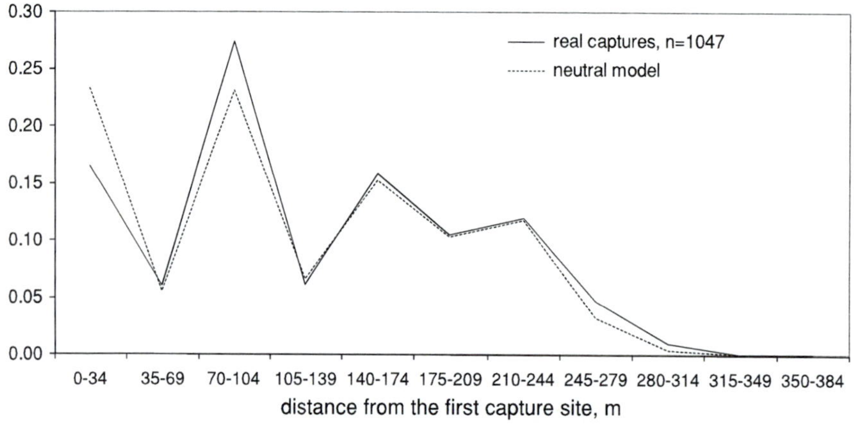

Fig. 6.4 The frequency distribution of distances between the capture locations of Eurasian reed warblers in moult (i.e. during postfledging movements) in late summer on the Courish Spit and the neutral model. The difference is not significant (Wilcoxon matched pairs test, $z = 0.97$, $p = 0.33$)

Cape Rossitten used this area together. This pattern was typical of both pre-migratory and migratory periods.

These results for *Acrocephalus* warblers are at variance with the pattern reported by Bibby and Green (1981) from France and Portugal. There these birds reportedly occupied exclusive home ranges and at least occasionally even

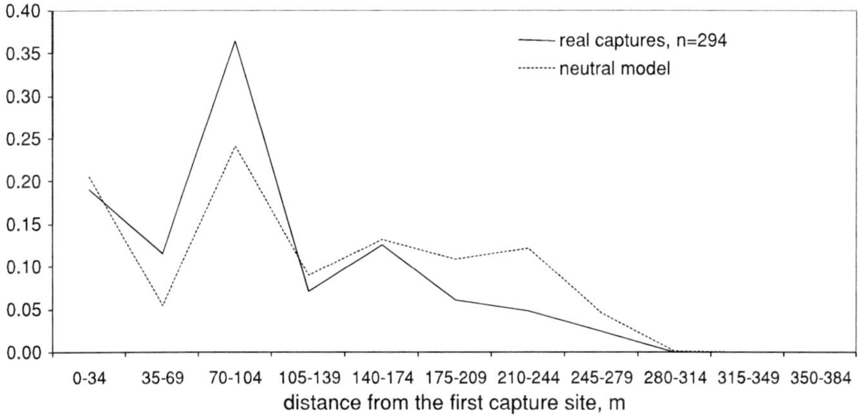

Fig. 6.5 The frequency distribution of distances between the capture sites of sedge warblers in autumn on the Courish Spit and the neutral model. The difference is not significant (Wilcoxon matched pairs test, $z = 0.77$, $p = 0.44$)

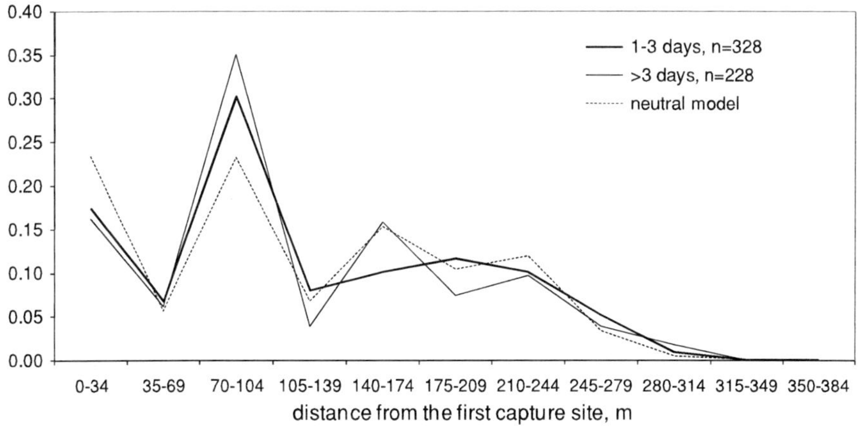

Fig. 6.6 The frequency distribution of distances between the capture sites of Eurasian reed warblers in summer and autumn on the Courish Spit and the neutral model. Birds that stopped for 1–3 days and for >3 days are treated separately. The difference is not significant (Friedman ANOVA, $F_{2,9} = 4.67$, $p > 0.05$)

defended them from conspecifics. These authors explain such spatial pattern by the diet of reed warblers that in Portugal were mainly feeding on dipterans that are easily disturbed, i.e. each individual should be carefully harvesting insects within its restricted home range and should not allow poaching by intruders. The authors showed that foraging efficiency expressed as the number of pecks per unit time was directly related to time elapsed since the last visit to the area. At the same time, sedge warblers mainly take immobile plum aphids, move broadly searching for their aggregations and may forage several metres from conspecifics. In the Baltic area both sedge and Eurasian reed warblers actively take plum aphids when they are abundant and both have to switch to alternative prey when they are scarce. In line with that, spatial behaviour of these two species did not differ, either. It may be noted that in Estonia plum aphids also form a substantial part of the Eurasian reed warbler diet (Mäll 1995). Thus, the spatial behaviour of *Acrocephalus* warblers may vary with their preferred diet and the distribution pattern of their prey. It may also be mentioned that during the breeding season, Eurasian reed warblers defend only a small area around their nest from conspecifics and share their foraging areas with neighbours (Catchpole 1972; Leisler and Schulze-Hagen 2011), i.e. their spatial behaviour is more similar to what we have seen during stopovers on the Baltic coast than to what Bibby and Green (1981) reported from western Europe.

Recapture analysis facilitated significant progress in our understanding of spatial behaviour of songbird migrants at stopover as compared with visual observations. However, this method has a very important basic limitation: detection probability of a bird depends on its mobility. To be captured in a mist-net, a migrant must move. The capture probability of a bird that remains stationary is close to zero.

If the movement pattern changes with time spent at stopover, as is the case in e.g. European robins and winter wrens (Fig. 6.2), detection probability changes, too. A European robin that occupied a restricted home range within the study area but relatively far from the lines of mist-nets has a very low detection probability. Incidentally, it shows that estimates of stopover duration based on recaptures (minimum stopover duration and stochastic models, Sect. 3.1) are seriously biased against the birds that make long stopovers, i.e. truncate the frequency distribution of stopover durations from the right.

This is well illustrated by recaptures of European robins in high mist-nets in Rybachy. As shown in Sect. 2.1.4, estimates based on recaptures in standard mist-nets are significantly biased low. The only way to overcome this difficulty is radio-tagging of migrants. We have studied the movements of radio-tagged European robins, pied flycatchers and sedge warblers on the Courish Spit (Chernetsov et al. 2004; Chernetsov and Mukhin 2006).

6.4 Spatial Behaviour at Stopovers: Radio-Tracking Data

6.4.1 Technical Parameters of Radio-Tracking and Data Analysis

We used LB-2 transmitters by Holohil Systems (Ontario, Canada) for tagging European robins and Micro-Pip transmitters by Biotrack (UK) for pied flycatchers and sedge warblers. Transmitters were fitted by Rappole harnesses (Rappole and Tipton 1991), the mass of tags with harness was 0.61 g for Holohil tags and 0.50 g for Biotrack tags. The added mass was 3.2–4.1% of European robin body mass, 3.4–4.1% of pied flycatcher and 3.4–4.3% of sedge warbler body mass, i.e. it always remained within 5% (Caccamise and Hedin 1985; Naef-Daenzer 1993). The life span of transmitters was at least 10 days for pied flycatchers, sedge warblers and European robins in spring and at least 21 days for European robins in autumn.

The location of birds was estimated by biangulation and triangulation. For each individual, one location per hour was taken between the onset of daytime activity (dawn) and evening civil twilight. The number of observations per individual per day varied between 11 and 17, depending on the duration of the daylight period. Locations were plotted on a digitised map of the study area. From sunset to dawn, all the birds were surveyed continuously from a stationary watch point 12 m above ground level.

We tested the locations for statistical independence by using the Schoener index (Swihart and Slade 1985). The data were not formally independent (i.e. consecutive locations were aggregated with a greater-than-chance probability); nevertheless, we assumed that our data could be used for the analysis of spatial distribution. We based our assumption on the empirical rule suggested by White and Garrott (1990), which states that if enough time has elapsed between two consecutive observations for an animal to move from one end of its home range to another, the observations in question may be considered statistically independent. In our study, at least 45 min elapsed between observations, during which each individual would have had ample time to move to any point in its stopover area. As for moving pied flycatchers, we refrained from estimating the size of their home ranges and considered only the linear scale of their movements.

When locating birds, every effort was made to approach them as closely as possible to minimise location error. We believe that in most cases we located their positions to the nearest 5 m. Home range area was estimated on the basis of all locations available as 95% kernel by Animal Movement Extension in ArcView (Hooge and Eichenlaub 2000). The estimated home range area increases with an increasing number of locations until that number reaches 40–50 (Lajda 2001); therefore, we did not estimate the home range area of birds with less than 38 locations. Due to this limitation, we only estimated the home range area for the entire stopover period and for the birds that stopped for at least 4 days ($n = 30$).

To estimate the aggregation of locations from birds that were followed during shorter periods of time, we used the linearity index as applied in Animal Movement Extension of ArcView (Hooge and Eichenlaub 2000); this is the linear distance passed (i.e. the distance between the initial and final locations) divided by cumulative distance between all successive locations. The maximum value of the linearity index is 1 (i.e. if a bird is moving along a straight line). This index may be calculated for a given time interval (e.g. the total observation period or a single day) and is a measure of area-restricted movement. The linearity index is reciprocal to the meander ratio (Williamson and Gray 1975) and was preferred to it due to the statistical properties of the linearity index.

Table 6.4 The number of European robins radio-tracked in spring and autumn 2002–2003 on the Courish Spit

Season	Tagged	Tracked since the first day of stopover	Tracked since the first days of stopover until departure	Fat	Lean
Spring					
2002	21	12	10	13	4
2003	30	30	29	16	14
Total	51	42	39	29	18
Autumn					
2002	29	25	24	10	19
2003	36	36	35	17	19
Total	65	61	59	27	38

We used the arbitrarily selected threshold of 0.10 as an indication that a bird occupied a defined stopover area; we assumed that birds showing linearity index values below this threshold settled in a home range. For comparison, Aborn and Moore (1997) found that the meander ratio for summer tanagers which settled at stopovers on the Gulf of Mexico coast averaged 4.8, which corresponds to a linearity index of 0.21.

6.4.2 Case Study: European Robin

We radio-tagged a total of 116 European robins (Table 6.4) in April–early May 2002 and 2003 and in September–October 2002 and 2003 (Chernetsov and Mukhin 2006). As we also aimed to obtain unbiased estimates of stopover duration (Sect. 2.3), we made every effort to capture and tag the birds on the first day after nocturnal arrival. On the Courish Spit, like in many other coastal areas, trapping numbers of many migrants, European robins included, strongly varies on the daily basis, with waves of new arrivals alternating with pauses when few migrants arrive (Dolnik 1975, 1981; Bolshakov 1981; Titov and Chernetsov 1999; Chernetsov and Titov 2000). Analysis of seniority based on time-inverted capture-mark-recapture models (Pradel 1996) showed that birds initially captured on the first day of a wave of arrivals must have arrived recently (Sect. 2.1.2, Table 2.1; Titov and Chernetsov 1999).

In 2003, all European robins ($n = 66$) were radio-tagged on the first day of a wave of new arrivals. In 2002, most birds were tagged on the first day ($n = 37$), but some ($n = 13$) were tagged on the second or third day. It should be mentioned that all the birds radio-tagged not immediately upon arrival were captured and ringed on the first day, i.e. their stopover duration was known with the same degree of accuracy. However, the data on their spatial behaviour on the first day(s) was lacking.

We managed to track 33 individuals in spring and 42 in autumn since arrival till departure. The number of locations taken varied between 6–92 in spring (in 1–6 days) and 4–172 in autumn (1–14 days). The linearity index varied between 0.008 in spring and 0.003 in autumn (very aggregated locations) and 0.65

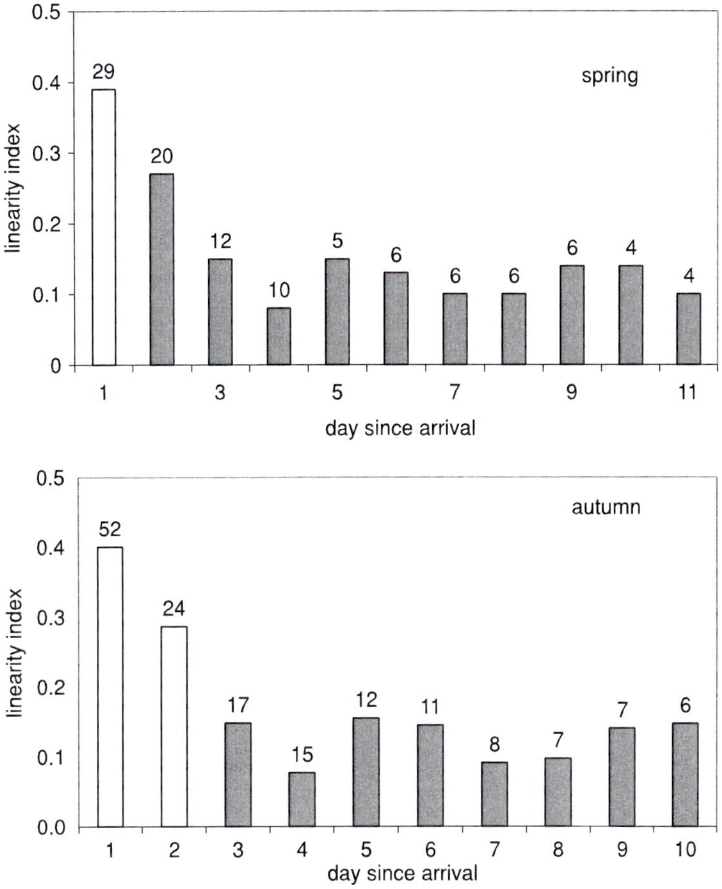

Fig. 6.7 The daily linearity index values of movements in European robins during their stopovers in spring and autumn on the Courish Spit. Sample sizes are shown above the histogram bars. The days with mean linearity index values significantly different from the remaining days (one-way ANOVA with post hoc tests) are shown by open bars

in spring and 0.93 in autumn. The latter case was a nearly straight line movement. In both seasons the linearity index was significantly inversely related to the number of locations taken (Spearman's rank correlation, $r_S = -0.69$ in spring and $r_S = -0.55$ in autumn) and to stopover duration in days ($r_S = -0.58$ in spring and $r_S = -0.56$ in autumn, all $p < 0.001$). The longer a European robin stayed the more aggregated were its locations, i.e. the more restricted was the area where it remained.

We also calculated linearity index values for each day of stopover (Fig. 6.7). The patterns in both seasons were similar: broader movements on the first day of stopover (in spring) or on the first two days (in autumn), after which locations became more aggregated, i.e. movements were more restricted. The difference in the mean linearity index between the days of stopover was significant in both seasons (one-way ANOVA, $F_{10,97} = 6.85$ in spring, $F_{9,149} = 6.69$ in autumn; $p < 0.001$). The linearity index was significantly greater on the first day of stopover in spring (Tukey's honestly significant difference test: $p < 0.008$) and on the first two days in autumn (Tukey's HSD test: $p < 0.001$). In autumn, the linearity index did not differ between the first and the second day of stopover (Tukey's HSD test: $p = 0.56$). Beginning with the third day of stopover, there was no significant between-day variation in the linearity index (post hoc tests; all $p > 0.05$).

Very similar results were obtained by Paxton et al. (2008) for Wilson's warblers during spring migratory stopover in the lower Colorado basin. Even though the authors claim that the values of linearity index decreased by 32% with every day of stopover, i.e. that decline was linear, a detailed inspection of Fig. 4 in their paper (Paxton et al. 2008) clearly shows that linearity index was significantly different only on the first day upon arrival, exactly like in European robins on the Courish Spit.

It should be emphasized that spatial behaviour on the first day upon arrival does not allow prediction whether the European robin is going to take-off after a one-day stopover, or it will remain for a longer period. The linearity index did not differ between birds continuing migration on the next night and those that remained for more than one day, either in spring (t-test, $t = 1.21$, $p = 0.20$, $n_1 = 14$, $n_2 = 15$) or in autumn ($t = 0.97$, $p = 0.34$, $n_1 = 28$, $n_2 = 27$). The pattern of movements on the last day before departure does not make it possible to predict take-off, either. As compared with the previous day, in 15 individuals the linearity index decreased and in 12 birds it increased (European robins that stopped over for at least three days, both seasons pooled). This proportion is not significantly different from 1 : 1 ratio ($\chi^2_{1,1} = 0.16$; $p = 0.69$).

In Figs. 6.8 and 6.9, representative maps of locations of several European robins at migratory stopovers are given. Locations of some individuals are situated entirely in a restricted area, whereas in others the proportion of locations was taken when they were still exploring the area (if taken at the beginning of stopover) or forays outside the home range. Such forays were more than once observed by me and also reported by Lajda (2001). It should be emphasized that in the cases when all locations are clustered in a small area, we may have missed the early exploration phase.

The home range area estimated across the whole stopover period by 95% kernel was on average 4,320 m^2 in spring and 3,562 m^2 in autumn (Table 6.5) and did not significantly differ between the seasons (t-test, $t = 0.94$, $p = 0.38$). In autumn, the area occupied by European robins on the Courish Spit was not significantly different from the home ranges on the coast of Lake Constance in SW Germany (Table 6.5, $t = 0.95$, $p = 0.35$). It should be emphasized that these estimates include the whole area used by the birds throughout the stopover period. In the advanced phase of a protracted stopover European robins may remain within a much smaller area, even by an order of magnitude.

On the other hand, Tsvey (2008), unlike us, made every effort to tag European robins not only in the first days after migratory arrivals, but within a few hours

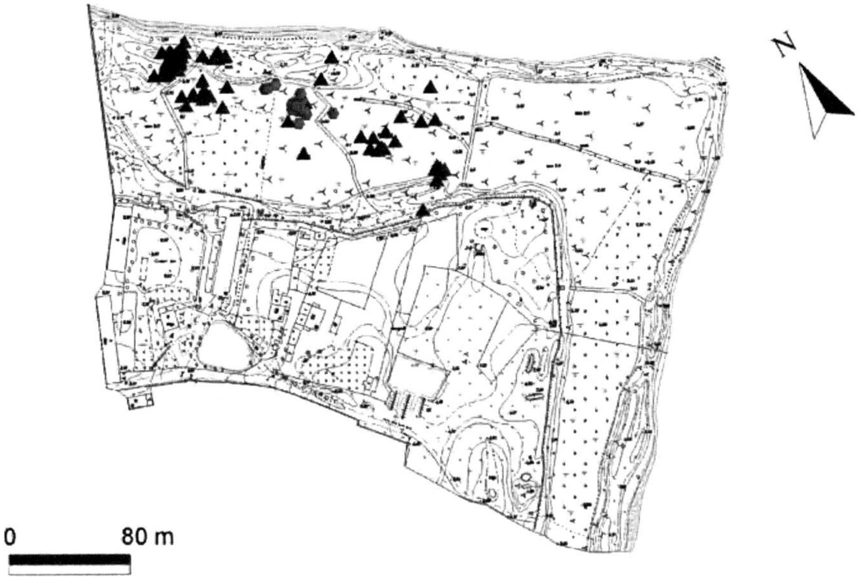

Fig. 6.8 Representative locations of several European robins during their migratory stopovers on the Courish Spit. Different symbols refer to different birds

Fig. 6.9 Representative locations of several European robins during their migratory stopovers on the Courish Spit. Different symbols refer to different birds

Table 6.5 Area of home ranges of European robins at stopovers on the Courish Spit (Eastern Baltic, our data) and in SW Germany (Laida 2001)

	Range, m^2	Mean, m^2	Median, m^2	SE	*n*
Spring, Rybachy	1,932–9,215	4,320	4,091	545	15
Autumn, Rybachy	1,060–10,083	3,562	2,801	598	15
Autumn, Germany	1,900–7,600	4,264	4,400	421	14

after this event. Not surprisingly, his data better cover the search/settling period, resulting in a larger total area utilised by European robins than our estimate. It should be also kept in mind that Tsvey (2008) did his research in another area on the Courish Spit which included pine plantations, mixed pine and birch forest and willow scrub on sand dunes. Most habitats at that site were suboptimal for stopover as compared with wetter habitats on Cape Rossitten, which may have contributed to longer exploratory movements before settling of European robins.

6.4.3 Spatial Behaviour of Other Songbird Migrants

In late May 2003, we radio-tagged 12 sedge warblers and tracked them until their migratory departure (Chernetsov et al. 2004). All these birds were tagged on the first day of a wave of arrivals, i.e. with a high probability soon after their actual arrival. Their stopover duration was 1–3 days, during which we obtained 9–40 locations, which is not sufficient for robust estimates of the home range area by kernel (Lajda 2001; Chernetsov and Mukhin 2006). Therefore, we were limited to estimating the linear scale of movements from the most distant location points.

The mean stopover duration of sedge warblers during spring migration was 1.6 days (SE = 0.23, median 1 day, n = 12). The maximum distance between locations varied between 44 and 335 m, on average 97 m (SE = 22.9, median 75 m, n = 12). A representative map of locations of two different individuals is shown in Fig. 6.10. Home ranges of sedge warblers, like in European robins, were apparently no defended territories. Home ranges of birds simultaneously present at the stopover site estimated by the minimum convex polygon method overlapped by 50% and more (Fig. 6.11). The linearity index of movements varied between 0.010 and 0.315 and was on average 0.147 (SE = 0.026, median 0.142, n = 12).

During the spring of 2003, we also radio-tagged seven pied flycatchers, all males (Table 5.6). Pied flycatchers showed no such pronounced waves of arrivals like European robins or sedge warblers, therefore in this case we are less confident that the birds were tagged immediately after arrival. Three birds stopped for one day, and one pied flycatcher resumed migration after a two-day stopover. Three more individuals were lost during their long-distance movements (Fig. 5.1), so we can only say that their minimum durations of stay were one, three, and four days, respectively. The distance between the most distant locations of individual pied

Fig. 6.10 Representative locations of several sedge warblers during spring migratory stopovers on the Courish Spit. Different symbols refer to different birds

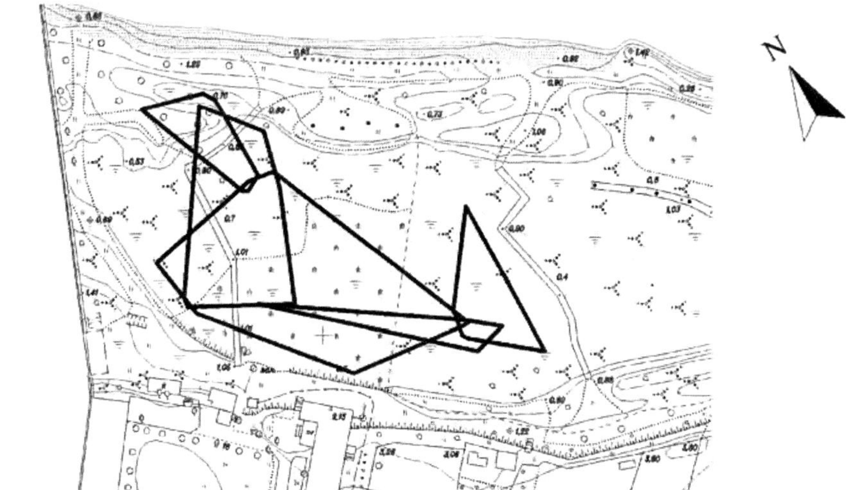

Fig. 6.11 Home rages of several sedge warblers simultaneously present at the stopover site on the Courish Spit shown by minimum convex polygon

flycatchers varied between 270 and 4,000 m (Table 5.6), on average 2,000 m (SE = 630, median 2,300 m, n = 7). Thus, pied flycatchers moved significantly longer distances than sedge warblers did (t-test, t = 3.73, $p < 0.002$). It should be emphasized that we underestimate the true distance of pied flycatchers' movements as we lost some of them during their daytime movements. The longest distance of daytime movements in this species that we recorded by recaptures was 10.7 km in 5 h (Sect. 6.2), but such long movements seem to be rather infrequent.

Linearity index of movements varied in pied flycatchers within a wider range than in sedge warblers: between 0.061 and 0.910, on average 0.421 (SE = 0.130, median 0.474, n = 7). Even though just seven pied flycatchers were tracked, it seems that they form two clusters: in three birds linearity index of their movements did not exceed 0.11, and in four it was greater than 0.45 (Table 5.6). Apparently, migrating pied flycatchers either remain within a small home range, like most other songbird nocturnal migrants do, or move for hundreds of metres or a few kilometres.

Summer tanagers moved on average for 328 m (76–1,166 m, SD = 277, n = 24) during spring stopovers on the northern coast of Gulf of Mexico (Aborn and Moore 1997). It should be however mentioned that in this study, radio-tracking was only continued for 4–11 h after tagging. Because of that, these data cannot be directly compared with our data on European robins, sedge warblers and pied flycatchers when every effort was made to track the birds until their migratory departure. The mean distance of movements of summer tanagers during the first day of stopover (383 m, SD = 277, n = 13) was not significantly different from the distance of movements of birds that had already spent at stopover at least one day (357 m, SD = 396, n = 5; t = 0.14, p = 0.90). The mean meander index of summer tanagers' movements was 3.6 (n = 24), which corresponds to linearity index of 0.28. The mean linearity index of movements on the first day of stopover was 0.34 (n = 13), i.e. nearly exactly equal to the value found in European robins in spring on the Courish Spit (0.33, n = 27).

Ovenbirds that stopped over in Prospect Park, Brooklyn, New York City, moved relatively broadly during the first four days of stopover and did not significantly decrease the linearity of their movements until the fifth day in spring, but settled already on the second day of stopover in autumn (Seewagen et al. 2010). The linearity index of their movements on the first day after arrival was ca. 0.25 in spring and ca. 0.30 in autumn, i.e. similar to the values observed in European robins and summer tanagers. Home ranges occupied by ovenbirds were on average 4.8 ha (SE = 1.1, n = 15) in spring and 3.7 ha (SE = 1.2, n = 13) in autumn (t-test, p = 0.48). These values were obtained by 95% kernel and are thus comparable with our data on the home range size in migrating European robins; ovenbirds' home ranges were larger by an order of magnitude.

Eurasian reed warblers tape-lured into the suboptimal habitat of willow scrub on sand dunes on the Courish Spit, where they normally do not occur on autumn migration, moved only short distances, 30–310 m (n = 10, Ktitorov et al. 2010). However, lean individuals moved longer distances than fat ones ($F_{1,8} = 8.2$, $r^2 = 0.51$, $p = 0.021$, n = 10). Birds that were released into the optimal habitat on

Cape Rossitten did not move longer than 200 m (one bird that moved 1,000 m into a non-reed habitat was excluded as an outlier; Ktitorov et al. 2010), i.e. behaved very similarly to sedge warblers (see above).

Garden warblers were radio-tracked during autumn migratory stopovers of Crete before crossing the Mediterranean and the Sahara (Fransson et al. 2008). Many birds remained in the same area during protracted (>12 days) stopovers, whereas others moved for 2–4 km after spending 2–3 days at stopover, i.e. apparently already after the search/settling period. One bird made two such relatively long-distance movements during its 14-day stopover. Thus, the spatial behaviour of garden warblers on Crete was a kind of compromise between what we observed on the Courish Spit in European robins and pied flycatchers.

6.5 Is Spatial Behaviour Condition-Dependent?

6.5.1 Radio-Tracking Data

An obvious candidate for a predictor of within-species variation in spatial behaviour is fuel stores. Our data on radio-tracking of European robins did not suggest any clear difference between spatial patterns of fat and lean individuals (Chernetsov and Mukhin 2006). The home-range area of birds that arrived fat (spring: 4,101 m^2, SE = 493, n = 5; autumn: 2,970 m^2, SE = 518, n = 6) and those that arrived lean (spring: 4,683 m^2, SE = 976, n = 8; autumn; 3,957 m^2, SE = 939, n = 9) did not differ either in spring (t-test, t = 0.44, p = 0.67) or in autumn (t = 0.80, p = 0.44).

In spring, the linearity index did not differ between birds that arrived lean and those that arrived fat on either the first day of stopover (fat: 0.34, SE = 0.039, n = 16; lean: 0.32, SE = 0.059, n = 11; median test: $\chi^2 = 0.30$, p = 0.58) or on the second day (fat: 0.18, SE = 0.037, n = 11; lean: 0.15, SE = 0.040, n = 6; median test: $\chi^2 = 0.03$, p = 0.86). In autumn, the pattern was similar: on the first day of stopover, linearity index was 0.44 (SE = 0.059, n = 22) in fat robins and 0.37 (SE = 0.37, n = 30) in lean individuals (median test: $\chi^2 = 1.26$, p = 0.26); on the second day, the respective figures were 0.21 (SE = 0.063, n = 10) and 0.34 (SE = 0.062, n = 14; median test: $\chi^2 = 2.74$, p = 0.10).

Apparently, both lean and fat European robins can show various spatial patterns in the first days after arrival. In both seasons the birds in both groups move more broadly on the first 1–2 days upon arrival, and occupy smaller home ranges in the subsequent days. Tracking Wilson's warblers in Arizona in spring also did not show any relationship between the spatial patterns and condition of the birds (Paxton et al. 2008).

On the other hand, other studies did show condition-related stopover movements. Exploratory movements of European robins in another area on the Courish Spit were weakly condition-related: lean individuals moved on average longer distances before they settled than fatter birds did (Tsvey 2008; Ktitorov 2012). Eurasian reed warblers also showed an inverse relationship between fuel load and scale of movements in both the suboptimal and the optimal habitat (Ktitorov et al. 2010). It should be however mentioned that all Eurasian reed warblers moved short distances, up to 200–300 m.

The same pattern was reported for summer tanagers (Moore and Aborn 2000) and ovenbirds (Buler 2006). Caged thrushes of four species (wood thrushes, veeries *Catharus fuscescens*, Swainson's thrushes and grey-cheeked thrushes *Catharus minimus*) in spring after crossing the Gulf of Mexico showed condition-related activity: lean individuals showed stronger restlessness than did fatter birds (Wang and Moore 1993). A similar pattern was found in free-living Palaearctic-African migrants that made stopovers in the Sahara: fat individuals stopped over in the desert, they were content with any shade, and did not move. Leaner birds tried to find optimal stopover habitats and more often occurred in the oases (Biebach 1985; Biebach et al. 1986).

6.5.2 Analysis of Recaptures of European Robins

Apart from radio-tracking, interesting information on fuel stores related spatial behaviour may be extracted from the analysis of recaptures. Titov (1999b) has elegantly demonstrated that during autumn migration, European robins that carried medium and large fuel stores [score 3 and more after Kaiser scale (Kaiser 1993)] were less often captured during the second day of their stopover than lean birds (no visible subcutaneous fat). Among birds whose minimum stopover duration exceeded two days, more than 40% of lean individuals were captured on the second day. Among fat robins, this proportion was below 10% (t-test for proportions, $t = 7.9$, $p < 0.05$). All these birds were present in the study area because they were captured afterwards. The difference in capture probability of fat and lean birds most likely was due to the greater mobility of the latter group.

Furthermore, Titov (1999b) showed that fatter European robins were significantly more active in the evening (since the last hour before sunset) than their lean conspecifics (Table 6.6). This difference was most apparent on the first day of stopover. In the subsequent days it was slightly reduced, but remained highly significant (Table 6.7). It is worth noting that the increased evening trapping rate of fat European robins was mainly achieved by captures (i.e. activity of the birds) in open habitats, reedbeds (Table 6.8). In willow scrub the proportion of captures in the last hour before sunset and in the evening twilight was not condition-related after the first day of stopover ($\chi^2_{1,1} = 2.19$, $p = 0.14$).

Apparently European robins with medium and large fuel stores move to open habitats in the evening, on the first day after arrival and in the subsequent days. As fat birds are generally ready to continue migration and make a nocturnal migratory flight, the possibility should be kept in mind that these movements do not carry any foraging function but are related to preparation for nocturnal flights. It has been suggested that at least some nocturnal songbird migrants calibrate their magnetic

Table 6.6 Daily pattern of captures of lean and fat European robins in autumn on the Courish Spit (from Titov 1999b, with alterations)

Fat score	Proportion			n
	Morning	Day	Evening	
0	0.346	0.575	0.079	1,187
>2	0.224	0.520	0.256	246
χ^2, significance	13.97; $p = 0.0002$	2.52; $p = 0.11$	65.45; $p < 0.0001$	

Table 6.7 Daily pattern of captures of lean and fat European robins in autumn on the Courish Spit depending of the day of stopover (from Titov 1999b)

Fat score	Stopover day	Proportion			n
		Morning	Day	Evening	
0	First	0.359	0.553	0.088	409
	Subsequent	0.339	0.586	0.075	778
	χ^2, significance	0.48; $p = 0.49$	1.23; $p = 0.27$	0.67; $p = 0.41$	
> 2	First	0.252	0.443	0.305	131
	Subsequent	0.191	0.609	0.200	115
	χ^2, significance	1.30; $p = 0.25$	6.76; $p = 0.009$	3.57; $p = 0.059$	

Table 6.8 Captures of fat European robins (fat score >2) in respect to habitat during stopovers on the Courish Spit after the day of arrival (from Titov 1999b)

	Proportion			n
	Morning	Day	Evening	
Reed	0.262	0.405	0.333	42
Willow scrub	0.251	0.726	0.123	73
χ^2, significance	0.03; $p = 0.86$	11.55; $p = 0.0007$	7.35; $p = 0.007$	

compass from twilight cues before each migratory flight (Cochran et al. 2004; Muheim et al. 2006, 2007). It seems logical that they move towards open areas from which they get a better view of the setting Sun and of the sky with its polarisation pattern. Therefore, it should be kept in mind that even though most diurnal movements of nocturnal migrants are habitat- and foraging-related, some specific movements may have another function.

6.6 Spatial Behaviour of Songbird Migrants at Stopover and Spatial Distribution of Food

Unfortunately, we are not aware of experimental studies that would show a relationship between spatial behaviour of migrants at stopovers and spatial distribution of their food. However, quite a number of indirect evidence suggests that such a relationship exists.

One of passerine species with a very pronounced tendency to occupy restricted temporary home ranges at stopovers is the European robin. This species forages during migration mainly on myriapods, spiders, coleopterans, ants, parasitic hymenopterans (Titov 2000; Chernetsov and Titov 2003). Observations of foraging European robins show that they mainly collect their food from the ground or from the trunks and large branches of trees and bushes low above the ground. Most invertebrates identified in the European robin diet usually occur in the ground layer. Within the respective habitats (scrub, forest with the shrubs layer) distribution of the food of this species is spatially relatively uniform and temporally predictable.

Another species which was reported to occupy restricted home ranges is the Eurasian reed warbler at stopovers in Portugal and France (Bibby and Green 1981). The authors explained this pattern by foraging on easily disturbed (and evenly distributed—N.C.) prey, mainly on dipterans. Our data from the Eastern Baltic do not confirm either this spatial pattern in this species (Sect. 6.3.2) or the diet (Chernetsov and Manukyan 1999a, b). In the Baltic area Eurasian reed and sedge warblers (Mäll 1995; Chernetsov and Manukyan 2000) and marsh warblers (Chernetsov 1999) readily take plum aphids that may form large aggregations but shows a much clumped distribution (Chernetsov 1998a). It may be mentioned that the famous paper by Bibby and Green (1981) often mentioned in textbooks and reviews (e.g. Bairlein 1996; Leisler and Schulze-Hagen 2011) is based on one year of study and on unclear methods of studying spatial behaviour and diet of migrants.

Our data on rather broad movements of pied flycatchers at stopover on the Courish Spit do not agree with the reports from autumn passage in Portugal where the birds of this species occupy small home ranges and even defend temporary territories (Bibby and Green 1980). However, it cannot be ruled out that the pattern of distribution of flying insects that form the basis of pied flycatcher diet on migration (to a greater degree than during breeding or in winter quarters; Alatalo and Alatalo 1979; Salewski 1999; Salewski et al. 2002, 2006; Chernetsov et al. 2004) is very different in September on the Iberian Peninsula and in May in the Eastern Baltic. In the Baltic area aggregations of flying insects occurred locally in the most insolated and wind-protected areas. The main goal of broad movements was apparently the search for such local concentrations of food. It is conceivable that in early autumn in Portugal the distribution of flying insects was less clumped.

Temporary territories on migratory stopovers have also been reported in rufous hummingbirds (Carpenter et al. 1993b). In these birds some individuals defended territories with a certain number of flowers (an important resource for these nectar-feeders), whereas some hummingbirds behaved as non-territorial roamers. It remains unknown whether these roamers were forced to use this territorial strategy (e.g. because their inferior social status did not allow them to occupy a territory), or this type of behaviour was optimal under a certain combination of conditions, e.g. because they did not need to invest energy and time into territorial behaviour.

Our comparative data on other songbird migrants does not allow us to rule out this possibility *a priori*.

Blackcaps switch to fruit-eating during autumn migration (Berthold 1976; Bairlein 1990; Eggers 2000). On the Courish Spit the main fruit taken by blackcaps is common elder (Zelenova 2001). Elder bushes are distributed in a clumped fashion; at the same time each group of elder bushes is a too abundant source of food to defend from the competitors: such behaviour is not justified by possible gains. Therefore, blackcaps roam broadly during migratory stopovers and overlap their home ranges with conspecifics (Chernetsov 2002). Garden warblers that also utilise fruits during autumn migration, in the experiment explored greater areas than did sedentary Sardinian warblers *Sylvia melanocephala momus* (Mettke-Hoffmann and Gwinner 2004).

All songbird migrants studied so far that move across larger areas during their migratory stopovers and not only do not defend territories but even do not remain within small home ranges (Eurasian reed and sedge warblers, pied flycatcher, blackcap) utilise food which is non-uniformly distributed in space. Taken together, this comparative data makes it possible to suggest that relatively uniformly distributed resources promote occupation of small home ranges, whereas utilisation of clumped food triggers broad movements in search of food aggregations. Birds that find locally abundant food do not defend these patches from conspecifics or heterospecific competitors. Our field data support the prediction of the theoretical model: animals that use non-uniformly distributed locally abundant food sources should forage in groups (and thus do not defend the resource), whereas species that utilise uniformly distributed food should forage solo (Clark and Mangel 1986).

Certainly, in real life the situation is much more complicated by the impact of food availability and the number of competitors, i.e. by the severity of competition. Unfortunately, very few studies have directly shown the presence of competition between migrants at stopovers (Moore and Wang 1991; Kelly et al. 2002; Ottich and Dierschke 2003; Salewski et al. 2007). The reason for this is probably the difficulty to formally demonstrate competition in field-based studies rather than the rarity of competition in the wild. However, in spite of many factors that may influence the situation and obscure the pattern, the basic principle seems to be that uniform distribution of food causes more or less uniform distribution of the birds through occupation of home ranges or territories. It should be noted that a migratory stopover is a simple situation from the viewpoint of optimisation criteria: the birds should maximise their instant rate of fuel gain while minimising the risk of predation (see Chap. 4). They do not need to select a nest site (which may modify or bias habitat requirements as compared with purely foraging-related habitat selection), they do not need to care about the future food availability when foraging demands of their offspring are high, whereas habitat productivity may change with the season. Therefore spatial and habitat behaviour of passerine migrants at stopovers may be a naturally simplified model that helps to understand the organisation of their spatial behaviour during other parts of the annual cycle.

References

Aborn DA, Moore FR (1997) Pattern of movement by summer tanagers (*Piranga rubra*) during migratory stopover: a telemetry study. Behaviour 134:1077–1100

Agatsuma R, Ramenofsky M (2006) Migratory behaviour of captive white-crowned sparrows, *Zonotrichia leucophrys gambelii*, differs during autumn and spring migration. Behaviour 143:1219–1240

Alatalo RV, Alatalo RH (1979) Resource partitioning among a flycatcher guild in Finland. Oikos 33:46–54

Bairlein F (1981) Ökosystemanalyse der Rastplätze von Zugvögeln. Ökol Vögel (Ecol Birds) 3:7–137

Bairlein F (1990) Nutrition and food selection in migratory birds. In: Gwinner E (ed) Bird migration. Springer, Berlin

Bairlein F (1996) Ökologie der Vögel. G. Fischer, Stuttgart, Jena, Lübeck, Ulm

Bastian A (1992) Mobilität von Kleinvögeln in einem suddeutschen Rastgebiet während der Wegzugperiode. Ökol Vögel (Ecol Birds) 12:121–163

Begon M, Townsend CR, Harper JL (2006) Ecology: from individuals to ecosystems, 4th edn. Blackwell, Oxford

Berthold P (1976) Animalische und vegetabilische Ernährung omnivorer Singvögel: Nahrungsbevorzugung, Jahresperiodik der Nahrungswahl, physiologische und ökologische Bedeutung. J Ornithol 117:145–209

Bibby CJ, Green RE (1980) Foraging behaviour of migrant pied flycatchers, *Ficedula hypoleuca*, on temporary territories. J Anim Ecol 49:507–521

Bibby CJ, Green RE (1981) Autumn migration strategies of reed and sedge warblers. Ornis Scand 12:1–12

Biebach H (1985) Sahara stopover in migratory flycatchers: fat and food affect the time program. Experientia 41:695–697

Biebach H, Friedrich W, Heine G (1986) Interaction of body mass, fat, foraging and stopover period in trans-Sahara migrating passerine birds. Oecologia 69:370–379

Bolshakov CV (1977) Izuchenie nochnykh migratsiy ptits [metodicheski aspekt] (Study of nocturnal bird migration [methodological aspect]). In: Ilyichev VD (ed) Metody izucheniya migratsiy ptits (Methods of studying bird migration). Nauka, Moscow

Bolshakov CV (1981) Rekonstruktsiya polnoy kartiny nochnogo prolyota i effektivnost' obnaruzheniya ego raznymi metodami (Reconstruction of the total picture of nocturnal passage and effectiveness of several methods of its estimation). Proc Zool Inst 104:95–123

Buler JJ (2006) Understanding of habitat use by land birds during migration along the Mississippi Gulf coast using scale-dependent approach. Dissertation, University of Southern Mississippi

Busse P (1995) New technique of a field study of directional preferences of night passerine migrants. Ring 17:51–75

Busse P (2000) Bird station manual. SEEN and University of Gdańsk, Gdańsk

Caccamise DF, Hedin RF (1985) An aerodynamic basis for selecting transmitter loads in birds. Wilson Bull 97:306–318

Carpenter FL, Paton DC, Hixon MA (1983) Weight gain and adjustment of feeding territory size in migrant hummingbirds. Proc Natl Acad Sci USA 80:7259–7263

Carpenter FL, Hixon MA, Russel RW, Paton DC, Temeles EJ (1993a) Interference asymmetries among age-classes of rufous hummingbirds during migratory stopover. Behav Ecol Sociobiol 33:297–304

Carpenter FL, Hixon MA, Temeles EJ, Russel RW, Paton DC (1993b) Exploitative compensation by subordinate age-classes of migrant rufous hummingbirds. Behav Ecol Sociobiol 33:305–312

Catchpole CK (1972) A comparative study of territory in the reed warbler (*Acrocephalus scirpaceus*) and sedge warbler (*A. schoenobaenus*). J Zool 166:213–231

Chernetsov N (1998a) Habitat distribution during the post-breeding and post-fledging period in the reed warbler *Acrocephalus scirpaceus* and sedge warbler *A. schoenobaenus* depends on food abundance. Ornis Svecica 8:77–82

Chernetsov N (1998b) On the direction of postnesting dispersal in the sedge warbler (*Acrocephalus schoenobaenus*) (Passeriformes, Sylviidae). Russ J Zool 2:626–628

Chernetsov N (1998c) Stopover length and weight change in juvenile Reed Warblers *Acrocephalus scirpaceus* in autumn in the Eastern Baltic. Avian Ecol Behav 1:68–75

Chernetsov NS (1999) Migratsionnye strategii kamyshevok *Acrocephalus* spp. v predelakh Evropy (Migratory strategies of *Acrocephalus* warblers within Europe). Dissertation, Zoological Institute RAS

Chernetsov N (2002) Spatial behaviour of first-year Blackcaps (*Sylvia atricapilla*) during the pre-migratory period and during autumn migratory stopovers. J Ornithol 143:424–429

Chernetsov N (2005) Spatial behavior of medium and long-distance migrants at stopovers studied by radio tracking. Annals New York Acad Sci 1046:242–252

Chernetsov N (2011) Daytime movements of nocturnal migrants at stopover between two nearby capture sites. J Ornithol 152:1007–1011

Chernetsov N, Manukyan A (1999a) Feeding strategy of Reed Warblers *Acrocephalus scirpaceus* on migration. Avian Ecol Behav 3:59–68

Chernetsov N, Manukyan A (1999b) Foraging strategies of the Sedge Warbler (*Acrocephalus schoenobaenus*) and the Reed Warbler (*A. scirpaceus*) on migration. Proc Zool Inst 281:101–106

Chernetsov N, Manukyan A (2000) Foraging strategy of the Sedge Warbler (*Acrocephalus schoenobaenus*) on migration. Vogelwarte 40:189–197

Chernetsov N, Mukhin A (2006) Spatial behaviour of European robins *Erithacus rubecula* during migratory stopovers: a telemetry study. Wilson J Ornithol 118:364–373

Chernetsov N, Titov N (2000) Design of a trapping station for studying migratory stopovers by capture-mark-recapture analysis. Avian Ecol Behav 5:27–33

Chernetsov N, Titov N (2001a) Movement patterns of European Reed Warblers *Acrocephalus scirpaceus* and Sedge Warblers *A. schoenobaenus* before and during autumn migration. Ardea 89:509–515

Chernetsov N, Titov N (2001b) Migratory stopovers of juvenile Blackcaps *Sylvia atricapilla* in autumn: stopover length, fuel deposition rate, and an attempt to predict departure body mass. Avian Ecol Behav 6:27–28

Chernetsov NS, Titov NV (2003) Pitanie i strategiya vesenney migratsii zaryanki, *Erithacus rubecula* (Aves, Turdidae), v yugo-vostochnoy Pribaltike (Foraging and spring migratory strategy of the robin *Erithacus rubecula* (Aves, Turdidae) in the southeastern Baltic Sea region). Zool Zhurnal 82:1525–1529

Chernetsov N, Mukhin A, Ktitorov P (2004) Contrasting spatial behaviour of two long-distance passerine migrants at spring stopovers. Avian Ecol Behav 12:53–61

Clark CW, Mangel M (1986) The evolutionary advantages of group foraging. Theor Popul Biol 30:45–75

Cochran WW, Mouritsen H, Wikelski M (2004) Migrating songbirds recalibrate their magnetic compass daily from twilight cues. Science 304:405–408

Dolnik VR (1975) Migratsionnoye sostoyanie ptits (Migratory disposition of birds). Nauka, Moscow

Dolnik VR (1981) Prognoz osennego dnevnogo proleta ptits na osnove dinamicheskoy modeli (Forecast of autumn daytime passage of birds on the basis of the dynamic model). Proc Zool Inst 104:133–143

Eggers S (2000) Compensatory frugivory in migratory *Sylvia* warblers: geographical responses to season length. J Avian Biol 31:63–74

Fedorov VA (1990) Barsuchok—*Acrocephalus schoenobaenus* (L.) [Sedge warbler—*Acrocephalus schoenobaenus* (L.)]. In: Rymkevich TA (ed) Linka vorobyinykh ptits severo-zapada SSSR (Moult of passerines of north-western USSR). Leningrad University Press, Leningrad

Fransson T, Barboutis C, Mellroth R, Akriotis T (2008) When and where to fuel before crossing the Sahara desert—extended stopover and migratory fuelling in first-year garden warblers *Sylvia borin*. J Avian Biol 39:133–138

Gwinner E (1996) Circadian and circannual programmes in avian migration. J Exp Biol 99:39–48

Hantge E, Schmidt-Koenig K (1958) Vom Herbstzug des Steinschmätzers (*Oenanthe oenanthe* L.) auf Wangerooge und Langeoog. J Ornithol 99:142–159

Hooge PN, Eichenlaub B (2000) Animal movement extension to Arcview. Ver. 2.0. Alaska Science Center, Biological Science Office, USGS, Anchorage

Kaiser A (1993) A new multi-category classification of subcutaneous fat deposits on song birds. J Field Ornithol 64:246–255

Kelly JF, Delay LS, Finch DM (2002) Density-dependent mass gain by Wilson's warblers during stopover. Auk 119:210–213

Kodric-Brown A, Brown JH (1978) Influence of economics, interspecific competition, and sexual dimorphism on territoriality in migrant rufous hummingbirds. Ecology 49:285–296

Ktitorov PS (2012) Vliyanie landshaftnykh usloviy na prostranstvennoe povedenie i izmenenie massy tela u migriruyuschikh vorobyinykh ptits (Influence of landscape conditions on spatial behaviour and body mass change in migrating passerines). Dissertation, Zoological Institute RAS

Ktitorov P, Tsvey A, Mukhin A (2010) The good and the bad stopover: behaviours of migrant reed warblers at two contrasting sites. Behav Ecol Sociobiol 65:1135–1143

Lajda M (2001) Telemetrische Untersuchung zum Rastverhalten des Rotkehlchens (*Erithacus rubecula*) in Südwestdeutschland während des Herbstzuges. Universität Zürich, Diplomarbeit

Leisler B, Schulze-Hagen K (2011) The reed warblers. KNNV Publishing, Zeist

Mäll K (1995) Roolindude (*Acrocephalus*) rondestrateegiatest Eestis. Diploma thesis, University of Tartu

Mettke-Hofmann C, Gwinner E (2004) Differential assessment of environmental information in a migratory and a nonmigratory passerine. Anim Behav 68:1079–1086

Mills AM, Thurber BJ, Mackenzie SA, Taylor PD (2011) Passerines use nocturnal flight for landscape-scale movements during migratory stopover. Condor 113:597–607

Moore FR, Aborn DA (2000) Mechanisms of en route habitat selection: how do migrants make habitat decisions during stopover? Studies Avian Biol 20:34–42

Moore FR, Wang Y (1991) Evidence of food-based competition among passerine migrants during stopover. Behav Ecol Sociobiol 28:85–90

Muheim R, Phillips JB, Åkesson S (2006) Polarized light cues underlie compass calibration in migratory songbirds. Science 313:837–839

Muheim R, Åkesson S, Phillips JB (2007) Magnetic compass of migratory Savannah sparrows is calibrated by skylight polarization at sunrise and sunset. J Ornithol 148(Suppl 2):S485–S494

Naef-Daenzer B (1993) A new transmitter for small animals and enhanced methods of home range analysis. J Wildl Manag 57:680–689

Ottich I, Dierschke V (2003) Exploration of resources modulates stopover behaviour of passerine migrants. J Ornithol 144:307–316

Paxton KL, Van Riper CIII, O'Brien C (2008) Movements patterns and stopover ecology of Wilson's warblers during spring migration on the lower Colorado river in southwestern Arizona. Condor 110:672–681

Pradel R (1996) Utilization of capture–mark–recapture for the study of recruitment and population growth rate. Biometrics 52:703–709

Ramenofsky M, Agatsuma R, Barga M, Cameron R, Harm J, Landys M, Ramfar T (2003) Migratory behavior: new insights from captive studies. In: Berthold P, Gwinner E, Sonnenschein E (eds) Avian migration. Springer, Berlin

Rappole JH, Tipton AR (1991) New harness design for attachment of radio-transmitters to small passerines. J Field Ornithol 62:335–337

Rappole JH, Warner DW (1976) Relationships between behavior, physiology and weather in avian transients at a migration stopover site. Oecologia 26:193–212

Redfern CPF, Alker PJ (1996) Plumage development and post-juvenile moult in the sedge warbler *Acrocephalus schoenobaenus*. J Avian Biol 27:157–163

Salewski V (1999) Untersuchungen zur Überwinterungsökologie paläarktischer Singvögel in Westafrika unter besonderer Berücksichtigung der Wechselwirkungen zu residenten Arten. W&T Verlag, Berlin

Salewski V, Bairlein F, Leisler B (2002) Different wintering strategies of two Palearctic migrants in West Africa—a consequence of foraging strategies? Ibis 144:85–93

Salewski V, Bairlein F, Leisler B (2006) Paläarktische Zugvögel in Afrika—Konkurrenz mit tropischen Arten? Vogelwarte 44:1–15

Salewski V, Almasi B, Heuman A, Thoma M, Schlageter A (2007) Agonistic behaviour of Palaearctic passerine migrants at a stopover site suggests interference competition. Ostrich 78:349–355

Schlenker R (1980) Nahrungssuche in Zugrichtung bei Limikolen. Vogelwarte 30:337–338

Schmaljohann H, Becker PJJ, Karaardic H, Liechti F, Neaf-Daenzer B, Grande C (2011) Nocturnal exploratory flights, departure time, and direction in a migratory songbird. J Ornithol 152:439–452

Schmidt-Koenig K (1980) Das Rätsel des Vogelzuges. Hoffman & Campe, Hamburg

Seewagen CL, Slayton EJ, Guglielmo CG (2010) Passerine migrant stopover duration and spatial behaviour at an urban stopover site. Acta Oecologica 36:484–492

Shumakov ME (1985) Orientatsia nekotorykh vidov nochnykh migrantov, ostanavlivayuschikhsia osenyu v severnoy chasti Pamiro-Alaya [Alayskaya dolina] (Orientaton of several species of nocturnal migrants that stop over in autumn in the northern part of the Pamir-Alay [the Alay valley]). Proc Zool Inst 138:136–142

Swihart RE, Slade NA (1985) Testing for independence of observations in animal movements. Ecology 66:1176–1184

Taylor PD, Mackenzie SA, Thurber BG, Calvert AM, Mills AM, McGuire LP, Guglielmo CG (2011) Landscape movements of migratory birds and bats reveal and expanded scale of stopover. PLoS ONE 6:e27054

Titov N (1999a) Individual home ranges of Robins *Erithacus rubecula* at stopovers during autumn migration. Vogelwelt 120:237–242

Titov N (1999b) Fat level and temporal pattern of diurnal movements of Robins (*Erithacus rubecula*) at an autumn stopover site. Avian Ecol Behav 2:89–99

Titov N (1999c) Home ranges in two passerine nocturnal migrants at a stopover site in autumn. Avian Ecol Behav 3:69–78

Titov N (2000) Interaction between foraging strategy and autumn migratory strategy in the Robin *Erithacus rubecula*. Avian Ecol Behav 5:35–44

Titov NV, Chernetsov NS (1999) Stokhasticheskie modeli kak novy metod otsenki prodolzhitelnosti migratsionnykh ostanovok ptits (Stochastic models as a new method for estimating length of migratory stopovers in birds). Uspekhi Sovremennoi Biologii 119:396–403

Tsvey AL (2008) Strategii migratsii zaryanki (*Erithacus rubecula*) v vostochnoy Pribaltike (Migratory strategies of the European robin (*Erithacus rubecula*) in the eastern Baltic). Dissertation, Zoological Institute RAS

Wang Y, Moore FR (1993) Relation between migratory activity and energetic condition among thrushes (Turdinae) following passage across the Gulf of Mexico. Condor 95:934–943

White GC, Garrott RA (1990) Analysis of wildlife tracking data. Academic Press, San Diego

Williamson P, Gray L (1975) Foraging behavior of the starling (*Sturnus vulgaris*) in Maryland. Condor 77:84–89

Wiltschko W, Höck H (1972) Orientation behavior of night-migrating birds (European robins) during late afternoon and early morning tests. Wilson Bull 82:149–163

Zehtindjiev P (1989) Kharakteristika yavleniy godovogo tsikla kamyshevok (*Acrocephalus* spp.) yuzhnogo Prichernomorya Bolgarii i ikh geograficheskaya izmenchivost' (Characteristics of the events of the annual cycle of reed warblers (*Acrocephalus* spp.) of the southern Black Sea coast in Bulgaria and their geographic variation). Diploma thesis, Leningrad State University

Zelenova N (2001) Weight gain and diet changes in young blackcaps (*Sylvia atricapilla*). Ring 23:179–189

Chapter 7
Temporal Pattern and Energy Cost of Migratory Flight

Abstract I claim that the currently widely accepted concept that migratory take-offs occur within a narrow time window soon after the sunset is not generally correct. Observation and radio-tracking data show that in a number of songbird migrants departures occur well into the night. The body mass and fuel stores of departing migrants vary broadly; a substantial proportion of birds take off with low fuel stores that do not allow them to fly throughout the night. Some migrants perform reverse migration during the night; however, others probably make short flights in the seasonally appropriate migratory direction. Our field data on the body mass loss during the flights support the recent wind tunnel measurements that suggest that migratory flight in long-distance migrants is significantly less expensive than was hitherto assumed: some 6–7 times basal metabolic rate (BMR), not 10–12 times BMR.

7.1 Time of Nocturnal Migratory Departures

7.1.1 Current Concept of Time Window of Departures

The current concept of the temporal schedule of nocturnal migratory flights of passerines is mainly based on surveillance radar studies (Drury and Nisbet 1964; Casement 1966; Parslow 1969; Gauthreaux 1971; Alerstam 1976). This concept assumes that departures of practically all nocturnal migrants occur in the period between 45 min after sunset and the end of evening nautical twilight. It is believed that migrants fly during 2–4 h and start to cease flight after midnight (Moore 1987; Kerlinger and Moore 1989). This pattern is suggested to be rather widespread between different species of songbird migrants; deviations from it are explained by landscape features of the parts of migratory routes, e.g. the necessity to cross water bodies (Martin 1990). The main reason for take-offs at the beginning of the night is

N. Chernetsov, *Passerine Migration*, DOI: 10.1007/978-3-642-29020-6_7,

the necessity to use different orientation cues (Emlen 1975, 1980; Moore 1987; Able 1982, 1989, 1993; Sandberg 1991) and to calibrate them (Cochran et al. 2004; Muheim et al. 2006, 2007). Quiescence period between the daytime activity and nocturnal flights (Palmgren 1949; Ramenofsky et al. 2003, 2008; Agatsuma and Ramenofsky 2006) occurs during evening twilight and may be related to making the decision to take off on a particular night and tackling orientation problems (Emlen 1980). This period is apparently typical not only of purely nocturnal migrants, but also of birds with the mixed diel rhythm of activity. For instance, fieldfares *Turdus pilaris* that migrate both during the day and at night, obligatorily interrupt their diurnal migratory flight at least 30 min before sunset and therefore make a break before setting off for nocturnal flights (Bolshakov 1992).

7.1.2 Visual Observations and Capture Data that Do Not Fit the Current Concept

The concept described above is predominant in bird migration research and mentioned in the reviews, including the most recent ones (Alerstam 1990; Berthold 2001; Newton 2008). However, the results of several studies based on visual observations (Bolshakov and Rezvyi 1998; Bolshakov and Bulyuk 1999, 2001) and the available radio-tracking data show that some birds may take off much later in the night (Cochran et al. 1967, 2004; Åkesson et al. 1996a, 2001; Bolshakov et al. 2007). This data makes it possible to question the concept of the narrow window of nocturnal departures and to suggest that take-off time in nocturnal passerine migrants may vary between species, latitudes and seasons.

During visual observations of spring migratory departures of European robins in horizontal searchlight beams, take-offs were recorded until the tenth hour after sunset, with median departure time of 152 min, i.e. 2.5 h, after sunset (Bolshakov and Bulyuk 1999). Before the end of nautical twilight, when migratory departures should cease following the concept of the narrow departure window, only 26% of European robins started their flights ($n = 156$).

In autumn, the situation was even more different from the commonly accepted view. Nocturnal take-offs of European robins also continued well into the tenth hour after sunset, the median departure time was 283 min after sunset, and only 4.6% of birds started flight in the first two hours of the night ($n = 741$, Bolshakov and Bulyuk 2001). Four and a half hours after sunset, when nocturnal migrants were supposed to be ceasing flight, one-half of European robins have not yet departed. The median departure time of song thrushes was 172 min after sunset, that of redwings, 250 min, and that of blackbirds, 386 min after sunset (Table 7.1). Song thrushes took off significantly earlier in the first half of the autumn migratory season (median take-off time 172 min after sunset) than in the second half (201 min; Kolmogorov-Smirnov test, $\lambda = 3.32$, $p < 0.001$; Bolshakov et al. 2002b), i.e. they started flights later into the night when nights were longer. Long-distance migrants (warblers and flycatchers) initiated their migratory flights

Table 7.1 Time of nocturnal departures of thrushes on the Courish Spit in autumn 1990–1995 (from Bolshakov et al. 2002b)

	Season of mass passage	Duration of night, h	n	Period of departure, hours after sunset	Median departure time, min after sunset	% departures in the first 2 h after sunset
Song thrush	10 September–12 October	10.9–13.3	745	1–10	172	30.2
Redwing	4–21 October	12.7–14.0	102	1–12	250	20.5
Blackbird	12–23 October	13.3–14.1	63	1–12	386	24.7

between the first and the eighth hours after sunset, with the median departure time of the pooled group 152 min (n = 752, Bolshakov and Bulyuk 2001). Long-distance migrants also shifted their departures deeper into the night with the progress of autumn migratory season.

Apart from migratory departures observed in the searchlight beams, more protracted than usually assumed take-off period was apparent from captures of birds in mist-nets elevated above the canopy (Bolshakov et al. 2000, 2003a, b). Captures of Eurasian reed warblers and sedge warblers during spring migration on the Courish Spit occurred throughout the night (Figs. 7.1, 7.2). However, the birds previously ringed in the standardised trapping project, i.e. individuals that were for certain taking off and not ceasing flight or flying very low above the ground, were only captured in the interval between the first and the fourth hour into the night. Captures of warblers without rings clearly form two clusters: some individuals are captured in the first half of the night simultaneously with the ringed, i.e. with certainty taking off, birds, whereas others are trapped in the second half of the night when not a single ringed Eurasian reed or sedge warbler occurred (Figs. 7.1, 7.2). The mean body mass of birds without rings captured in the first half of the night was not significantly different from the body mass of the ringed conspecifics (t-test, $t = 0.16$, $p > 0.10$ for Eurasian reed warblers; $t = 1.10$, $p > 0.10$ for sedge warblers). Most probably, all or at least the bulk of Eurasian reed and sedge warblers captured in elevated mist-nets in the first half of the night without rings were actually taking off for migratory fights. Departure time as shown by captures varied in both species between 45 and 240 min after sunset; the median time was 84 min (n = 60) in Eurasian reed warblers and 94 min (n = 36) in sedge warblers. Take-offs of both *Acrocephalus* species were rather synchronised, with 83–85% of individuals starting their flights between 45 and 120 min into the night.

In marsh warblers the dots representing the temporal distribution of departures form a single cloud (Fig. 7.3). Therefore, it is not possible to say whether the unringed individuals were departing or arriving. For some reason, marsh warblers during spring migration in the Eastern Baltic may start nocturnal flights later during the night than reed or sedge warblers, emphasizing variation even between closely related species of migrants.

Captures of European robins during autumn migratory departures occurred within a broader interval (second-twelfth hour into the night, median 260 min,

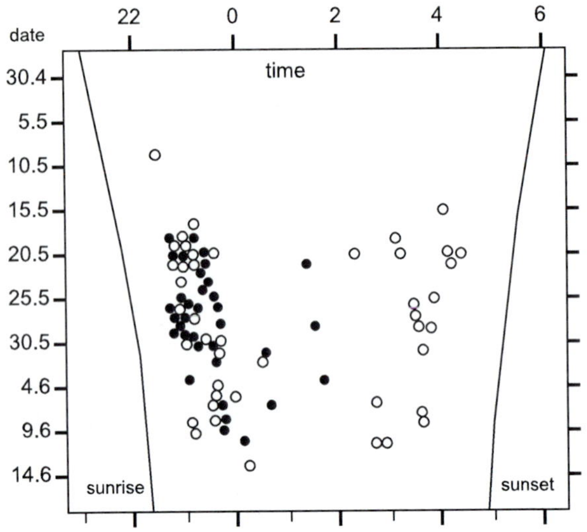

Fig. 7.1 The temporal distribution of nocturnal captures of Eurasian reed warblers in elevated mist-nets in spring on the Courish Spit. *Filled dots* represent the birds with rings (taking off); *open dots* are the birds without rings that may be taking off or ceasing flight. The *lines* show sunset and sunrise time

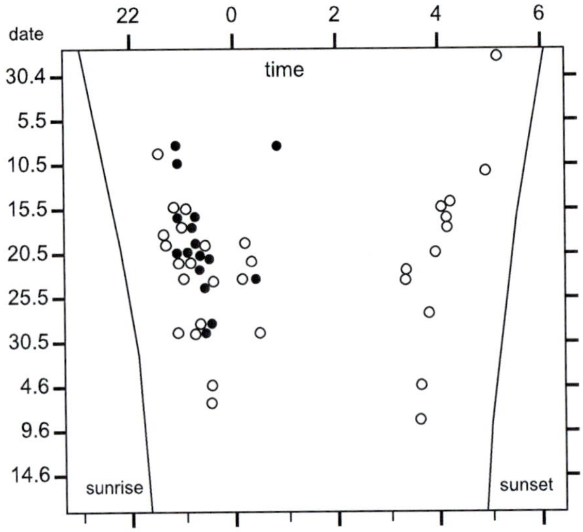

Fig. 7.2 The temporal distribution of nocturnal captures of sedge warblers in elevated mist-nets in spring on the Courish Spit. Symbols as in Fig. 7.1

n = 74) than in *Acrocephalus* warblers (Bulyuk and Tsvey 2006). Most birds (40%) took off in the second and third hours after sunset (Fig. 7.4). The proportion of early departures was higher later in the migratory season (Fig. 7.5). Bulyuk and Tsvey (2006) tested the difference in the mean time of departure by Mann-Whitney test and found no significant difference; however, the temporal patterns of departure did differ ($\chi^2_{1.5} = 14.5$, $p = 0.012$), probably influenced mainly by the duration of the night.

We can conclude that the common concept of the narrow window of nocturnal departures is more or less supported by the data on long-distance migrants,

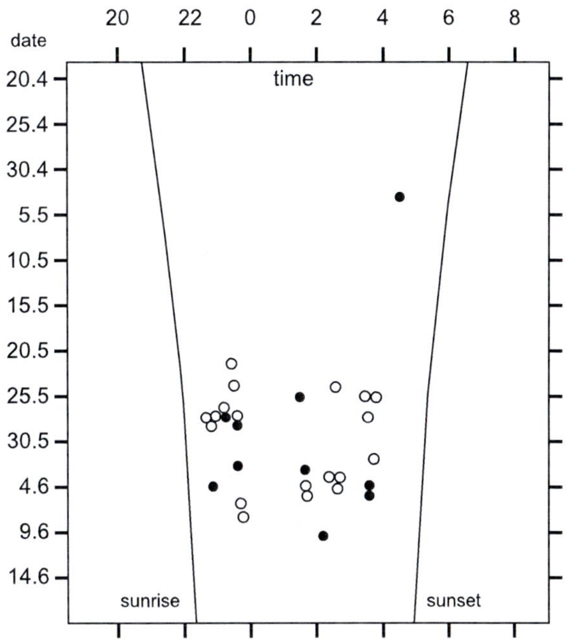

Fig. 7.3 The temporal distribution of nocturnal captures of marsh warblers in elevated mist-nets in spring on the Courish Spit. Symbols as in Fig. 7.1

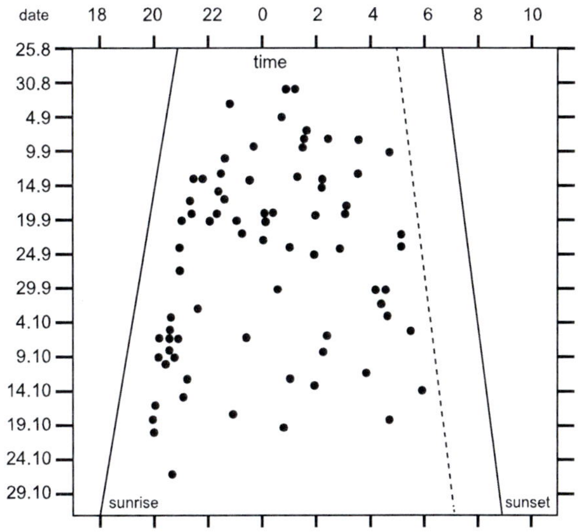

Fig. 7.4 The temporal distribution of nocturnal migratory departures of European robins in autumn on the Courish Spit as shown by captures of previously ringed birds in elevated mist-nets (from Bulyuk and Tsvey 2006)

especially when they migrate in spring under long days and short nights (Eurasian reed and sedge warblers migrate in the Baltic area in late May–early June when the Sun is above the horizon during 16–17 h). When long-distance migrants travel in autumn when the nights are longer, they take off later into the night. Take-offs of short- and medium-distance migrants: European robins, song thrushes, blackbirds,

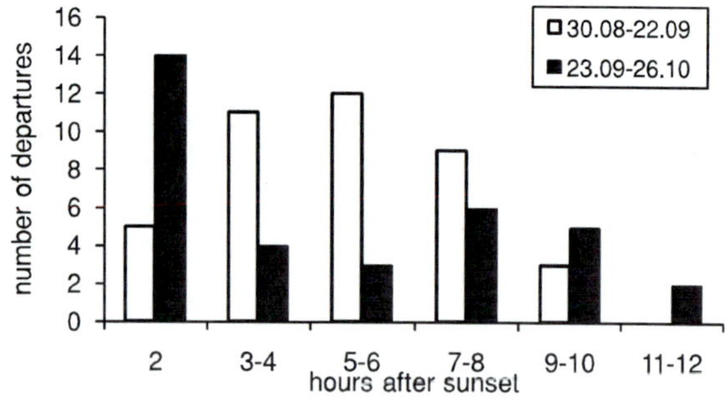

Fig. 7.5 The frequency distribution of nocturnal migratory departures of European robins on the Courish Spit as shown by captures in elevated mist-nets in the former and the latter half of autumn migratory season (from Bulyuk and Tsvey 2006)

fieldfares and goldcrests (Bolshakov and Bulyuk 1999, 2001) occurred throughout the night and did not fit into the concept of the narrow window of departures. It has been recently suggested that late take-offs were actually not true migratory departures but referred to exploratory flights or landscape-scale nocturnal movements (Mills et al. 2011). Such movements indeed occur in nocturnal songbird migrants, both during the migratory season (Schmaljohann et al. 2011) and during pre-migratory postfledging movements (Mukhin et al. 2005). However, I argue that late take-offs of e.g. European robins (Fig. 7.4) are simply too numerous to be explained by landscape-scale exploratory flights alone, even though such flights most probably do contribute to late-night departures.

7.1.3 Nocturnal Departures of Radio-Tagged Songbirds

The aforementioned data may be challenged: the searchlights that illuminated the area may have induced departures. Therefore we studied nocturnal take-offs in radio-tagged European robins (Bolshakov et al. 2007). Our data did not support the existence of the narrow time window of nocturnal departures during the evening nautical twilight, either in spring or in autumn. Moreover, take-offs of robins were not linked to any position of the Sun below the horizon (Figs. 7.6, 7.7). During evening twilight, i.e. when the Sun was between 0° and 18° below the horizon, only 14% departures occurred in autumn and 33% in spring. One-half of all departures took place more than 5 h after sunset in autumn and more than 3.5 h in spring, and 14% of birds in autumn and 7% in spring did not start flight until morning twilight.

The median departure time of European robins was 319 min after sunset in autumn (83–743 min, $n = 58$) and 208 min in spring (70–450 min, $n = 42$). The difference between the seasons was significant (Mann-Whitney test, $z = 3.69$, $p < 0.001$): in spring take-offs occurred on average 1.8 h sooner after sunset than in autumn. The position of the Sun also differed: during spring departures, the Sun was closer to the horizon than in autumn ($z = 3.49$, $p < 0.001$).

Thus, the radio-tagging data on the temporal schedule of nocturnal migratory departures were in a very good agreement with the data of visual observations

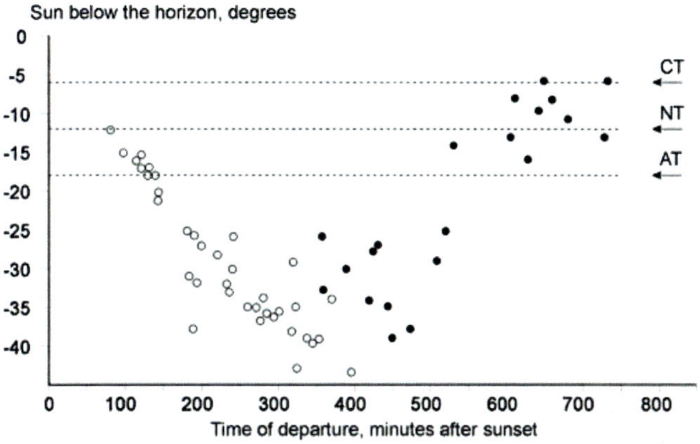

Fig. 7.6 Departure time of radio-tagged European robins in relation to elevation of the sun in the eastern Baltic in autumn. *Arrows* indicate the sun elevation in the end of civil (*CT*), nautical (*NT*) and astronomical (*AT*) twilight. *Open circles*: departure times when the sun was falling; *closed circles*: departure times when the sun had passed the lowest elevation of the night

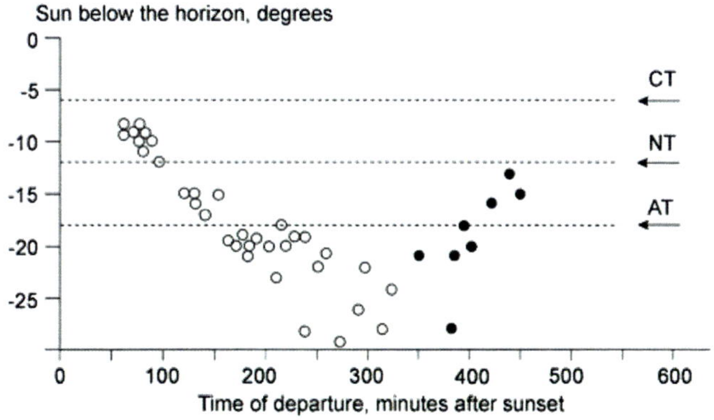

Fig. 7.7 Departure time of radio-tagged European robins in relation to elevation of the sun in the eastern Baltic in spring. Symbols as in Fig. 7.6

of take-offs and with nocturnal captures in the mist-nets elevated above the canopy. This agreement strongly suggests that migratory departures throughout the night are a natural phenomenon and not an artefact of the study method. It should be also emphasized that the probability of recording the departure of a radio-tagged songbird is independent of the weather (wind strength and direction, fog, visibility condition) that might influence visual recording or capture probability. Another benefit of radio-tracking is that when the birds take-off soon after tagging (within 1–2 days), their body mass is roughly known.

Apart from the data on departures of European robins, the temporal schedule of nocturnal departures was studied in Eurasian reed warblers in Falsterbo in the southernmost Sweden during autumn migration (Åkesson et al. 2001), and in the sedge warblers and pied flycatchers in spring on the Courish Spit (Bolshakov and Chernetsov 2004). Eurasian reed warblers initiated their flights between 32 and 634 min after sunset (median 128 min, n = 29). In this species, take-offs in autumn during on average longer nights occurred significantly later into the night than in spring, as indicated by captures at departure (median 84 min, Mann-Whitney test, $z = 3.68$, $p < 0.001$). Sedge warblers in spring under short nights took off 62–137 min after sunset (median 100 min, n = 12). All sedge warblers started nocturnal flights rather synchronously when the Sun was 7°–12° below the horizon. The median departure time of pied flycatchers was 145 min after sunset (41–317 min, n = 4). One pied flycatcher took off in the latter half of the night under overcast (317 min after sunset), but it aborted flight and landed one kilometre from the take-off site.

Mills et al. (2011) studied migratory departures of Swainson's and hermit thrushes in autumn on the northern coast of Lake Erie. They found that true migratory departures occurred at the beginning of the night (the median time was 0.7 h after the end of evening civil twilight [in their study area and during their study period, evening civil twilight ended on average 30 min after sunset]), whereas take-offs that started local and landscape-scale nocturnal movements occurred throughout the night with the median time of 6 h after sunset. These results are thus at variance with the results of other radio-tracking studies of nocturnal departures. At least as for spring departures of European robins from the Courish Spit (the median time 208 min, range 70–450 min, n = 42; Bolshakov et al. 2007), we can safely assume that a vast majority of them were indeed migratory departures and not local movements, because the birds were flying across the water of the Courish Lagoon with the nearest land ca. 30 km in front of them. The same was true of spring departures of sedge warblers and pied flycatchers from the Courish Spit (Bolshakov and Chernetsov 2004) and of autumn departures of Eurasian reed warblers from Falsterbo (Åkesson et al. 2001).

The mean departure time of radio-tagged northern wheatears during spring migration from Helgoland was 183 mm (SD = 66, median 176 min, n = 26), and exploratory flights of the same individuals in previous days occurred on average 173 min after sunset (SD = 71, n = 11; Schmaljohann et al. 2011). Thus, exploratory flights of northern wheatears that might be identical with local or landscape-scale flights of Mills et al. (2011) did not occur later into the night than actual departures, and take-offs of both kinds occurred well in the deep night, i.e. were not restricted to the evening twilight.

Of 18 summer tanagers that were tracked during spring migration after crossing the Gulf of Mexico on its northern coast, 14 individuals departed later than 4–5 h after sunset (Moore and Aborn 1996). Unfortunately, tracking was discontinued after 22–23 h local time, so that the exact take-off time of most individuals remained unknown. However, it is noteworthy that during a comparatively long night at 30° N in April, long-distance passerine migrants also started flights not at the beginning of the night. It seems that the time of departure is mainly governed by the duration of the night, with more birds starting migratory flights late under long nights, rather than by short vs. long-distance of migration.

It should be however mentioned that Cochran (1987) found a very restricted time window (9–13 min after the end of evening civil twilight) for migratory departures of one individual Swainson's thrush tracked during several subsequent flights. This is a very interesting observation, but the sample size of one precludes further speculations.

7.1.4 Impact of Weather on Departure Time

The departure time of European robins was not related to cloud cover in either season (Table 7.2). Some 40% of robins in autumn and ca. 21% in spring took off under strong cloud cover or overcast (cloud score 8–10). Similar results were obtained by radio-tracking for sedge warblers and pied flycatchers (Bolshakov and Chernetsov 2004). Stepwise multiple regression analysis of the weather factors vs. departure time of European robins captured in elevated mist-nets resulted in the model that explained only 8.5% of variation in departure time, i.e. was not satisfactory (Bulyuk and Tsvey 2006). However, when these authors analysed take-offs of robins that resumed migration after one-day stopover, multiple regression model explained 55% of variation in the dependent variable and included weather parameters related to cloud score and precipitation, and the synoptic weather situation. The latter parameter was strongly correlated with the tailwind component, i.e. with wind assistance for migratory flight. When the authors analysed separately European robins that departed after prolonged stopovers, no usable models with weather variables were obtained at all (Bulyuk and Tsvey 2006).

The authors suggest that this difference in response to weather condition in European robins that resume nocturnal flights after one-day and after longer stopovers suggests significant motivational and probably physiological variation between these two groups. Apparently, favourable weather condition (mainly good wind assistance) triggered the birds to continue their flights in the next night after arrival. The more favourable the weather conditions were, the less fuel robins needed to embark for a migratory flight: fuel stores at departure and tail wind component were inversely related ($r = -0.55$, $n = 27$, $p = 0.003$; Bulyuk and Tsvey 2006).

Schmaljohann et al. (2011) used the Sun elevation rather than departure time in their analyses of take-offs of northern wheatears from Helgoland. Neither wind speed nor direction significantly influenced departures in their study. Mills et al. (2011) found North American thrushes to make what they called 'true departures' earlier in the night under favourable winds. In southern Sweden Eurasian reed warblers also started nocturnal flights more eagerly under tailwinds (Åkesson et al. 2001). Wind assistance is an important factor that influences the readiness of songbird migrants to embark on a nocturnal flight (Richardson 1978, 1990; Švažas 1993; Åkesson and Hedenström 2000); however, some recent data suggest that it may be less decisive than often thought (Karlsson et al. 2011). As we can see, the impact of this factor of the timing of departures is much less obvious.

Table 7.2 The nocturnal departures of radio-tagged European robins from stopovers on the Courish Spit in relation to cloud cover

	Autumn		Spring	
	Cloud score at departure			
	0–7, mean 4.7	8–10, mean 9.5	0–7, mean 1.5	8–10, mean 9.2
n	15	15	33	9
Median date	6/7 Oct	6/7 Oct	21/22 Apr	19/20 Apr
Median departure time, min after sunset	290	271	204	250
Range, min after sunset	83–740	114–743	72–450	70–440
Percentage of departures with the Sun >18° below the horizon	80% (n = 11)	73% (n = 11)	54% (n = 12)	56% (n = 5)

7.2 Body Mass and Fuel Stores at Nocturnal Migratory Departure

Fuel load at migratory departure is a very important characteristic for understanding the relationship between the two main parameters of migration, stopovers and flights. Such data may be obtained either by captures at departure or by radio-tracking. In the latter case only migrants that resume flight after one-day stopovers yield useful information on their departure fuel loads, because during longer stopovers body mass may change considerably. It should be emphasized that routine captures at stopovers include birds just after arrival, in the process of refuelling and soon before take-off. The last capture of individual migrants does not generally occur immediately before their departure: in European robins on the Courish Spit, the time elapsed between the last capture in standard mist-nets and nocturnal departure could be up to 10 days, with the 'hidden' part of stopover up to 91%, on average 46% (see Sect. 2.1.4). Therefore, it is risky, to put it mildly, to assume that the body mass at the last recapture is representative of body mass at departure (e.g. Hall 1996).

Not much data is available on body mass of migrants at take-off. The body mass of Eurasian reed warblers departing for spring migratory flights on the Courish Spit varied between 11.2 and 14.6 g (mean 12.8 g, SD = 0.65, n = 60, Fig. 7.8). Their fuel stores calculated as the difference between body mass and estimated size-specific lean body mass varied between 0 and 3.35 g (mean 1.29 g, SD = 0.62), and the fuel load was on average 11.2% of their lean body mass (SD = 5.44). I assume the energy density of fuel stores to be 21.6 $kJ \cdot g^{-1}$. This value was obtained in the experiment with the thrush nightingale flying in the wind tunnel for many hours (Klaassen et al. 2000) and is close to the value 25.3 $kJ \cdot g^{-1}$, obtained by Dolnik and Gavrilov (1973) on the eve of studies of flight cost in passerines. This value of energy density of fuel stores is much lower than the energy density of pure fat that was used in the calculations of e.g. Bolshakov et al.

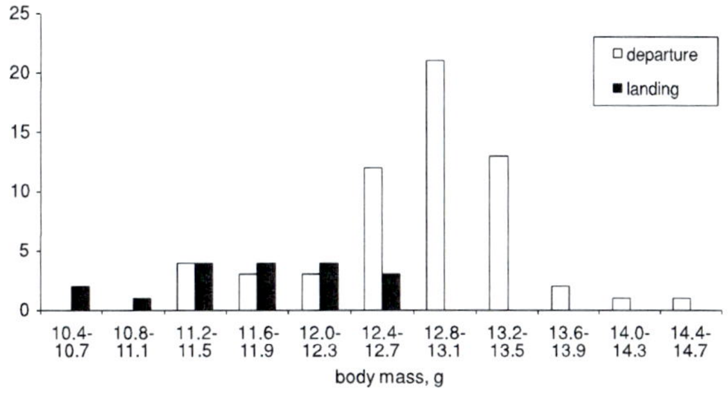

Fig. 7.8 The frequency distribution of body masses in Eurasian reed warblers at migratory departure and at arrival in spring from captures in elevated mist-nets on the Courish Spit

(2003a, b). The energy cost of flying during 1 h in the thrush nightingale is 6.876 kJ (Klaassen et al. 2000), which makes it possible to estimate flight cost in the Eurasian reed warbler at 3.65 kJ h^{-1} from the assumption that flight cost is proportional to body mass raised to the power of 0.665 (Eq. 5.8 from Dolnik 1995). Therefore, 1 g of fuel allows a Eurasian reed warbler to fly during 5.9 h, and fuel costs of birds that took off for nocturnal flights from the Courish Spit allowed them to fly on average during 7.6 h (range 0–19.8 h).

It should be emphasized that these estimates of potential flight duration are considerably higher than our estimates published earlier (Bolshakov et al. 2003a) because here I used different flight cost values. These values are based on new experimental data that I believe to be more reliable and realistic. On the other hand, these estimates are close to the values based on the rule by Delingat et al. (2008) that assumes losing 1% of body mass per hour of migratory flight: they would arrive at 9.9 h of flight on average.

It should be also noted that our estimates of fuel loads at take-off (and thus estimates of potential flight range based on them) are rather conservative. It has been recently shown that migrating passerines, including *Acrocephalus* warblers, may have very low body mass when crossing the Sahara, much lower than their usually assumed lean body mass, and remain alive, recover and continue migration (Salewski et al. 2010). Carcass analysis of Blyth's reed warblers captured on migration in Central Asia showed that birds without visible subcutaneous fat actually still had on average 0.5 g of fat (SD = 0.1, n = 13); in paddyfield warblers *Acrocephalus agricola* the respective value was 0.4 g (SD = 0.1, n = 13; Yablonkevich 1987). In common yellowthroats *Geothlypis trichas*, ovenbirds and Swainson's thrushes extractable lipid stores in birds with the no visible subcutaneous fat were 9.8–19.7% of their dry body mass (Seewagen 2008). It should be however kept in mind that a proportion of the lipids is an emergency margin that can only be used for flight under *force majeure* conditions, and a proportion is structural fat not available for metabolic purposes at all (Blem 1990; Biebach 1996).

The mean departure body mass of sedge warblers in spring on the Courish Spit was 13.7 g (11.6–17.0 g, SD = 1.17, n = 36). Their fuel stores were 0.14–4.62 g (mean 2.05 g, SD = 1.06), and the fuel load was on average 17.6% of their lean body mass (SD = 8.91). Fuel stores of departing sedge warblers allowed them to fly during 0.8–27.3 h, on average 12.1 h. These estimates are also higher than those given in the original paper (Bolshakov et al. 2003b), for the same reason.

The body mass of marsh warblers at spring migratory departures from the Courish Spit (only previously ringed birds) varied between 11.8 and 13.9 g (mean 13.0 g, SD = 0.66, n = 9). Unlike Eurasian reed or sedge warblers, ringed apparently departing marsh warblers were captured throughout the night, since the first hour after sunset until the penultimate hour of the night (Fig. 7.3). These birds were extracted from the nets 65–60 min before sunrise, i.e. in the morning twilight. Therefore, we cannot be certain that their captures did refer to migratory departures; these birds may have already started their normal daytime activities. The mean body mass of marsh warblers with certainty captured in the night and thus definitely at take-off, was 12.6–13.9 g (mean 13.2 g, SD = 0.25, n = 6). Their fuel stores weighed 1.6–3.2 g and allowed them to fly during 9.0–17.4 h; their fuel load was on average 23.1%.

It is worth noting that fuel mass and thus potential flight duration and range was significantly lower in Eurasian reed warblers than in sedge warblers (t-test, $t = 4.46$, $p < 0.001$) or marsh warblers ($t = 4.46$, $p < 0.001$). The reason for that might be that a proportion of sedge and marsh warblers were taking off from the Courish Spit for flights towards the north across the Baltic Sea (Chernetsov 1999), whereas the vast majority of Eurasian reed warblers were heading towards the north-east across land, as suggested by their ringing recaptures (Bolshakov et al. 2001, 2002a) and by the results of tests in Emlen funnels (Chernetsov et al. 2008).

The body mass of European robins captured at departure from the Courish Spit (both season pooled) varied between 13.6 and 23.2 g, on average 16.2 g (SD = 1.31, median 16.1 g, n = 78). The fuel mass was 0–7.99 g, on average 1.08 g (SD = 1.17, median 0.87), and the fuel load 6.2% (SD = 5.85). This amount of fuel allowed European robins to fly during 5.1 h on average (assuming flight power in this species to be 4.57 kJ h^{-1}).

The mean body mass of European reed warblers captured at take-off in spring on the Courish Spit (12.8 g) exceeded the mean mass of all birds trapped during their daytime movements in the area (12.15 g, SD = 0.84, n = 1,211, t-test, $t = 5.82$, $p < 0.001$). However, if only birds captured in the evening, after 18:00 local time, are included (i.e. when their body mass is comparable with the mass of individuals departing at the beginning of the night), the difference is no longer significant: 12.88 g, SD = 0.91, n = 47, $t = 0.60$, $p = 0.55$. The same pattern can be observed in the sedge warbler: the mean body mass of individuals captured at take-off (13.7 g) was higher than the mean body mass of sedge warblers trapped throughout the day (12.70 g, SD = 1.06, n = 652, $t = 5.35$, $p < 0.001$) but not significantly different from the mass of birds captured in the evening (13.72, SD = 0.97, n = 21, $t = 0.14$, $p = 0.87$).

In European robins captured in autumn during daytime movements on the Courish Spit throughout the day the mean body mass was very similar to the mass of departing birds (16.10 g, SD = 1.17, n = 23,420, $t = 0.86$, $p = 0.39$). The birds captured in the evening were on average even heavier than the departing individuals (16.77 g, SD = 1.15, n = 1,419, $t = 4.10$, $p < 0.001$). The reason for that was probably body mass loss during the night, so that European robins that took off in the latter half of the night had lost mass since sunset (Zimin 2003).

It may be concluded that body mass of several songbird migrants during migratory departure is not significantly higher than the mean mass of all conspecifics mist-netted during stopover. This is especially interesting if we keep in mind the fact that at least in some species (e.g. in the European robin) heavier individuals may have smaller capture probability and thus be underrepresented in captures, biasing the mean body mass estimate low.

7.3 Departures of Lean Birds and Nocturnal Reverse Flights

As we have just seen, among departing songbird migrants there are rather lean individuals whose body mass hardly exceeds the size-specific lean body mass. Even keeping it in mind that our estimates of fuel stores of migrants are rather conservative (see above), we still face the fact that migrants with very low fuel stores sometimes take-off for nocturnal flights. One explanation is that lean individuals perform nocturnal reverse migration looking for good stopover areas (Åkesson et al. 1996b; Åkesson 1999). In an earlier paper I expressed doubt that reverse nocturnal migration existed, because the arguments allowed alternative interpretations (Chernetsov 2006). The difference in body mass of Eurasian reed warblers that took off in autumn in the migratory and the reverse directions from Falsterbo (southernmost Sweden) were at the edge of statistical significance (means 14.7 g, $n = 21$ and 13.6 g, $n = 6$; one-way ANOVA: $F_{1,25} = 3.78$, $p = 0.063$; Åkesson et al. 2002). However, the body mass of Eurasian reed warblers radio-tagged in this study was 12.3–17.0 g (Åkesson et al. 2001), i.e. no really lean birds were included at all. Taking the general south–south-westerly direction of autumn migration into account together with the fact that the nearest land (the Zealand island) in the south–west is only 25 km away, even the leanest of Eurasian reed warblers in this study could have easily crossed this 'ecological barrier'. Second, the accuracy of detection of vanishing direction of small passerines tagged by BD-2B transmitters by Holohil Systems (Åkesson et al. 2001, 2002) raises doubts. Our experience of tracking small songbird migrants at departure (small *Acrocephalus* warbler or European robin size) does not allow us to be sure that the flying bird is always lost after it has selected flight direction, and not e.g. during circling around.

Recaptures of migrants north of the ringing site in autumn that mainly occur with lean individuals (Åkesson 1999) are even less convincing. Most of such events happened in August and could refer to birds that had not yet commenced migration and performed postfledging movements, including nocturnal ones (Bulyuk et al. 2000; Mukhin 2004; Mukhin et al. 2005, 2009). Postfledging nocturnal movements may also explain the radar data on northward movements in south Sweden (Zehnder et al. 2002) and in Israel (Komenda-Zehnder et al. 2002). However, southward radar tracks recorded in Israel in spring most probably refer to birds that perform reverse migratory flights.

Only very recently the data became available which proves the existence of short-range nocturnal movements not generally directed towards the goal of migration. Swainson's thrushes with low fuel stores (fat score not exceeding 2 in 6-grade scale) do take flights towards the north from Fort Morgan Peninsula on the northern coast of Gulf of Mexico (Jaci Smolinsky, ''personal communication".). Nocturnal flights towards the north were clearly recorded by automatic receiving units; reception range was at least 25 km. The birds were flying towards the north across Mobile Bay; the distance of such flights remained unknown but the minimum estimate is ca. 20 km.

At least before crossing the Gulf of Mexico (the minimum distance to Yucatan Peninsula is ca. 1,000 km) Swainson's thrushes indeed sometimes fly in the direction opposite the migratory one, probably looking for better stopover sites. It seems that when the birds do not have to cross such a serious ecological barrier, they may make landscape-scale flights in any direction, not necessarily the reverse one (Mills et al. 2011; Taylor 2011). It may be more correct not to call such relatively small-scale (tens of kilometres) nocturnal flights 'exploratory' (Schmaljohann et al. 2011) or 'landscape-scale' flights (Mills et al. 2011) rather than reverse migration, because movement in the reverse direction may be just a special case of landscape-related movements not oriented in respect to the goal of migration.

7.4 Is Departure Time Condition-Dependent?

The hypothesis that departure time is condition-dependent was first suggested by Moore and Aborn (1996) on the basis of their data of take-off time of 18 radio-tagged summer tanagers during spring migration in the southern United States. This idea looked coherent and replaced the concept of departure threshold of fuel stores (Biebach 1985) when it became apparent that some of departing nocturnal migrants had low fuel stores (Moore and Aborn 1996). This study was, unfortunately, based on a relatively small sample size, but most importantly, the exact time of migratory departures was not determined in 14 birds out of 18 (see Sect. 7.1.3).

Two studies were devoted to detailed analysis of the relationship between fuel load of nocturnal migrants and their departure time (Bulyuk and Tsvey 2006; Bolshakov et al. 2007). Both studies analysed take-offs of European robins, which is not surprising: as we have seen, the period of departures in this species is not restricted to a narrow window, but continues nearly throughout the night. Bulyuk and Tsvey (2006) analysed captures of departing robins in elevated mist-nets, and Bolshakov et al. (2007) investigated take-offs of radio-tagged individuals.

The authors of the former paper found no relationship between departure time and condition index of migrants (standardised for the beginning of the night; Fig. 7.9). Analysis of covariation showed no link between the arrival body condition, stopover duration and efficiency and time of migratory departure (ANCOVA, $F_{3.70} = 0.627$, $p = 0.60$). On the basis of these data Bulyuk and Tsvey (2006) rejected the hypothesis that the time of nocturnal departures is

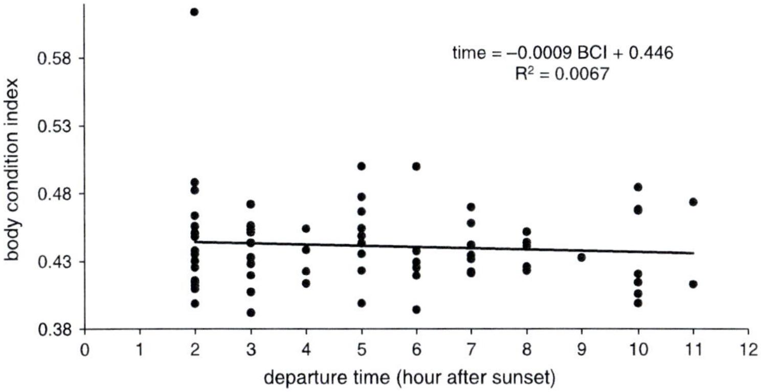

Fig. 7.9 Relationship between take-off time in European robins captured in elevated mist-nets in autumn on the Courish Spit, and their body condition index at the beginning of the departure night (from Bulyuk and Tsvey 2006).

Table 7.3 The departure time of European robins from stopovers on the Courish Spit in relation to their arrival fuel load (from Bolshakov et al. 2007)

	Arrival fuel load, g (mean ± SD)	n	Departure time, min after sunset (median, range)
Autumn	Low (0.44 ± 0.41)	21	429 (144–740)
	High (2.89 ± 0.66)	18	310 (122–743)
Spring	Low (0.53 ± 0.49)	6	264 (123–450)
	High (2.60 ± 0.44)	16	207 (79–440)

controlled by fuel stores of migrants to any significant degree, at least in European robins during autumn migration.

Bolshakov et al. (2007) analysed take-offs of radio-tagged European robins, therefore these authors could only be relatively certain about departure fuel loads of birds that resumed migration soon after capture and tagging, i.e. after 1–2 day stopovers. In both seasons, fat birds departed somewhat sooner after sunset than individuals with small fuel stores (on average by 57 min in spring and by 119 min in autumn), but this difference was not statistically significant (Mann-Whitney test, $z = 0.44$, $p = 0.66$ and $z = 1.52$, $p = 0.13$, respectively; Table 7.3).

In both seasons, European robins that resumed nocturnal migration after long stopovers (>2 days) and had a possibility to improve their energy status, took off on average sooner after sunset than lean birds after short (1–2 days) stopovers. The inverse relationship between departure time and stopover duration was significant in both seasons ($r = -0.43$, $p < 0.001$, $n = 58$ in autumn; $r = -0.50$, $p < 0.001$, $n = 42$ in spring). In autumn, this relationship was significant only in birds that had arrived lean (and probably gained much; $r = -0.52$, $p = 0.02$, $n = 33$), but not in individuals with large arrival fuel load ($r = -0.34$, $p = 0.10$, $n = 25$). In spring, both groups showed significant relationships ($r = -0.52$, $p = 0.021$, $n = 19$ and $r = -0.48$, $p = 0.019$, $n = 23$). It should be however kept in mind that in all cases the correlations reached significance because of individuals that made long stopovers (>4 days among initially lean birds and >10 days among robins with large arrival fuel loads).

In northern wheatears that took off in spring from Helgoland to fly across the North Sea departure time was related to their fuel load, with 18% of the variation in the timing of departure explained by this factor (Schmaljohann and Naef-Daenzer 2011).

It seems that fuel stores may have some effect on departure time of European robins and probably other nocturnal passerine migrants, but this is not the only factor that governs the time of nocturnal take-off. Birds that make long stopovers and probably significantly increase their fuel stores tend to take-off relatively soon after sunset. Individuals that resume nocturnal migration after 1 or 2 days of stopover may take-off early during the night even with seriously depleted fuel stores, if flight conditions (first of all wind direction) remain favourable. Apparently, when departures are well synchronised and mainly occur soon after sunset (like in long-distance migrants migrating under short nights, especially in spring in the north), there is not much variation in departure time anyway, so the energy status of migrants may have very restricted impact on their take-off time.

7.5 Time of Ceasing Migratory Flight

It is widely assumed that nocturnal migrants take off soon after sunset and generally cease flight in the middle of the night, at least when they are not crossing any ecological barriers (Moore 1987; Kerlinger and Moore 1989). This concept is probably not correct, as shown for instance by our tape-luring data (Tables 5.1, 5.2, 5.3, 5.4). If we compare the trapping figures during the deep night (Tables 5.1, 5.3) vs. the trapping figures during the morning twilight (Tables 5.2, 5.4), we can see that in autumn, apart from Eurasian reed and sedge warblers whose song was played back, only garden warblers and blackcaps were captured in any considerable numbers, whereas in spring, hardly any birds apart from tape-lured Eurasian reed warblers were captured before twilight. *Acrocephalus* warblers were attracted by song playback and probably induced to land, whereas playing non-attractive redwing song resulted in only three captures during the deep night. Thus, our capture data support the suggestion of Bolshakov (1981) based on visual observations that landing after nocturnal migratory flights mainly occurs in morning twilight.

The exact data on the time of ceasing migratory flights is much scarcer than the data on departure time. The reason for this is methodological (newly arriving birds are typically unmarked), and the situation is further complicated by departures throughout the night in some species and situations (see above). An unmarked European robin captured in elevated mist-nets may be either arriving or departing, and it is usually not possible to discriminate between these possibilities (Fig. 7.4).

Our data on migratory arrivals of birds captured in elevated mist-nets without song playback shows that Eurasian reed and sedge warblers in spring start to cease flights in the latter half of the night since the beginning of the third hour before sunset (Figs. 7.1, 7.2). In marsh warblers arrivals seems to overlap with departures, so it is not possible to say when these birds start to land (Fig. 7.3). This is even truer of European robins that continue to take-off until soon before sunrise.

Very interesting data were reported by Bulyuk (2006). He captured ten individuals of long-distance migrants (six Eurasian reed warblers, three marsh warblers and one lesser whitethroat) in elevated mist-nets when they were completing their last nocturnal flight during spring migration and landing into the site where they had been marked earlier. Of these 10 individuals, seven were ceasing flight during nautical twilight 120–60 min before sunrise. One Eurasian reed warbler and one marsh warbler were captured in the fourth hour before sunrise in astronomical twilight, and the only lesser whitethroat was captured during the deep night, i.e. when the Sun was more than 18° below the horizon. It is most probable that quite a few migrants (or even most of them) land in the last hour before sunrise when it is no longer dark. However, such birds cannot be distinguished from individuals that just start their daytime routine activity.

On the other hand, a proportion of migrants apparently land in complete darkness, as shown by the tape-luring data. When *Turdus* thrushes migrate with the following winds, at least 75% of individuals cease flights at the end of the night. However, under strong opposing winds the proportion of thrushes that stop flights at the beginning or middle of the night may be as high as 45–60% (Bolshakov et al. 2002b). Many direct observations of birds that enter the funnel traps in the darkness provide further evidence of landing before the end of the night (Shapoval 1981).

These data suggest that variation in the time of ceasing flight is considerable. This is further supported by the data of Cochran (1987) on the single radio-tracked Swainson's thrush that took off within a 5 min time range during several consecutive nights but landed within the interval of 170 min even when it did not encounter cold fronts.

It should be also kept in mind that this pattern (flight until the end of the night under favourable conditions, but occasional landfalls much earlier, due to inclement weather or other factors) is typical of the migration with a continuous stopover opportunity. Radar studies in the Sahara have shown that even though migration traffic rate across this major barrier gradually decreased at the end of the night, some Palaearctic migrants continue to fly across the desert throughout the day until the afternoon (Schmaljohann et al. 2007a, b). It should be however noted that species identification was generally not possible, therefore a proportion of daytime tracks could refer to daytime, not typical nocturnal migrants. Apparently, landbirds that cross large water bodies, like the Mediterranean or Gulf of Mexico, and encounter the sunrise over the water, have to fly until they reach land. Such daytime arrivals of nocturnal migrants have been reported. But typical nocturnal migrants that fly over land and do not encounter cold fronts or other inclement weather normally cease their flights in the latter half of the night, most often in morning twilight.

7.6 Arrival Body Mass

The body mass of migrants captured in the early morning on the first day of an influx of migrants mainly represents the arrival body mass. Such data is available from many trapping projects. The problem is that it is generally not possible to assume that individuals captured in the early morning hours are an unbiased sample of all migrants that have landed a couple of hours earlier. In the vast majority of projects migrants are trapped in mist-nets, i.e. the capture probability is strongly related to the mobility of migrants. This mobility may or may not be condition-dependent (Sect. 6.5), and generally we cannot assume that our samples are unbiased with respect to the body mass and fuel load of arriving migrants.

Captures in elevated mist-nets directly during migratory arrival (Bolshakov et al. 2000) are more likely to be an unbiased sample of arriving individuals. We have data on the body mass and fuel loads of Eurasian reed warblers and sedge warblers captured in elevated mist-nets when ceasing their migratory flights on the Courish Spit in spring. The body mass of Eurasian reed warblers captured at landing was 10.5–12.4 g, on average 11.7 g (SD = 0.67, n = 18, Bolshakov et al. 2003a). It was on average by 1.1 g smaller than body mass at migratory departure at the same site, the difference being significant (t-test, $t = 6.31$, $p < 0.001$). The fuel mass of landing individuals varied between 0 and 1.07 g, on average 0.36 g (SD = 0.35); the fuel load was on average 3.2% (cf. 11.2% at departure). The body mass of Eurasian reed warblers captured when completing their last spring migratory flight was 11.0–12.7 g, on average 11.9 g (SD = 0.60, n = 6; Bulyuk 2006). The body mass of individuals of this species tape-lured in spring in the latter half of the night was on average 11.9 g (range 10.1–14.0 g, SD = 0.75, n = 140). All these values are very close and probably represent a correct estimate of body mass and fuel load of Eurasian reed warblers completing their spring migratory flights in Europe.

The body mass of sedge warblers captured when ceasing flight varied between 11.7 and 15.8 g and was on average 13.1 g (SD = 1.41, n = 12; Bolshakov et al. 2003b). The mean fuel load of these birds was 11.1%, which is significantly higher than in Eurasian reed warblers ($t = 2.46$, $p = 0.03$). It was on average 0.6 g lower than the body mass at take-off, but this difference was not statistically significant ($t = 1.27$, $p > 0.10$). The mean body mass of tape-lured sedge warblers at the end of the night was 12.1 g (10.5–15.7, SD = 0.96, n = 46), which is significantly lower than the body mass of their conspecifics captured without song playback (t-test, $t = 2.84$, $p = 0.006$). It may suggest that lean sedge warblers (and very probably other migrants) are more likely to be attracted by song playback than their fatter conspecifics, and that the mean body mass in tape-luring samples is biased low.

7.7 Flight Power Estimates

The energy cost of horizontal steady-state flight expressed in the units of basal metabolic rate (BMR) is usually believed to be ca. 12 BMR (Dolnik 1995) or 10 BMR (Berthold 2001). This means that during flapping flight, the birds spend 10–12 times more energy per unit time than when resting. It is usually believed that because of the bioenergetic similarity of the birds the flight power to the resting power ratio remains roughly the same in many diverse avian species, despite the large range of BMR values (Dolnik 1995). Passerines have generally higher BMR and flight power than many non-passerines (Gavrilov 2011), but at least in songbirds energetic similarity is believed to be maintained.

However, direct measurements in the wind tunnel have shown that thrush nightingales have the flight power of 1.91 ± 0.07 W (Klaassen et al. 2000) which is equal to 6.876 kJ h^{-1}, or 6.0 BMR of this species. The BMR value of the thrush nightingale is calculated from another paper of the same authors (Lindström et al. 1999). It should be noted that wind tunnel studies, if anything, bias flight power estimates high because of non-uniform air stream (Masman and Klaassen 1987; Rayner 1994; Pennycuick et al. 1997); they should not bias estimates low. Therefore, direct measurements in a low-turbulence wind tunnel suggest that flight power in long-distance passerine migrants may be much lower than usually assumed.

Our data on arrival and departure fuel loads may be used to obtain an indirect estimate of flight costs. The mean difference between body mass at take-off and at arrival during spring migration on the Courish Spit was 1.1 g. Assuming the energy density of fuel of 21.6 kJ g^{-1} (Klaassen et al. 2000), this is an equivalent of 23.76 kJ. The mean duration of nocturnal migratory flights in this species in spring was estimated at 4.1 h (Bolshakov et al. 2003a), and it should be kept in mind that this value may be an underestimate, because we may have assumed too early mean time of ceasing flight. As already mentioned, a proportion of Eurasian reed warblers may have landed during the last hour before sunrise, when it was no longer possible to distinguish them from birds at stopover that have started their daily activities. Therefore, the estimate that flight power in this species is 5.80 kJ h^{-1} (from spending 23.76 kJ for flying during 4.1 h), or 1.61 W, may be slightly biased high.

I used the data on Eurasian reed warblers to obtain estimates, because in this species the estimates of body mass and fuel load at ceasing flight obtained by different methods are very similar and probably representative. In the sedge warbler the body mass of departing and arriving birds was not significantly different. The body mass of European robins captured in elevated mist-nets at departure (on average 16.2 g, Sect. 7.4) was significantly higher than the mean body mass of individuals captured during the first two morning hours (15.57 g, $SD = 0.98$, $n = 2{,}463$, $t = 5.70$, $p < 0.001$), but the difference was only 0.63 g. In European robins take-offs occurred throughout the night and broadly overlapped with arrivals, so the body mass of morning captures is the closest we can get to estimating arrival mass. The mean arrival time is not available, either, so there is no way to estimate the mean duration of flight in this species. Anyway, estimates based on mass loss during flight suggest flight power considerably lower than 10 BMR (Berthold 2001) or 12 BMR (Dolnik 1995).

As BMR of Eurasian reed warblers is close to 19.5 kJ day^{-1} (Bolshakov et al. 2003a), the flight power of 5.80 kJ h^{-1} is equivalent to the cost of migratory flight of 7.1 BMR. This is close to the estimate for the thrush nightingale in the wind tunnel study (6.876 kJ h^{-1}, or 6.0 BMR) and somewhat lower than the estimate for two free-flying *Catharus* thrushes (Swainson's thrush and hermit thrush). These thrushes spent 15.5 kJ h^{-1} in the natural migratory flight (Wikelski et al. 2003). BMR is 41.0 kJ day^{-1} in the hermit thrush and 40.2 kJ day^{-1} in the Swainson's thrush (Holmes and Sawyer 1975), on average 40.6 kJ day^{-1}. Thus, when flying, these thrushes were spending energy 9.2 times more rapidly than when resting.

It should be mentioned that our earlier estimate of flight power of sedge warblers based on body mass of two individuals captured in Estonia immediately after nocturnal flights from Finland (Bulyuk and Chernetsov 2000) is too high. We assumed the energy density of their fuel expenditure (1.0 and 1.2 g) to be equal to the energy density of pure fat (39 kJ g^{-1}), which is incorrect. Assuming energy density of fuel to be 21.6 kJ g^{-1} (Klaassen et al. 2000), flight cost is estimated at 4.91 and 9.26 kJ h^{-1}, which corresponds to 6.0 and 11.4 BMR. As speculated in the original publication (Bulyuk and Chernetsov 2000), the latter value may represent the increased cost of flying into the cold front, which that bird is known to have done.

Taken together, our flight cost estimates of Eurasian reed warblers (7.1 BMR) and sedge warblers (6.0 and 11.4 BMR) from mass loss during flight and estimates for the thrush nightingale flying in wind tunnel (6.0 BMR) and free-flying *Catharus* thrushes (9.2 BMR) are lower than 12 or even 10 BMR. Passerines adapted for long-distance flights seem to spend some 6–7 BMR when flying under calm air conditions (not in a cold front, when air turbulence may significantly increase the flight cost [Bulyuk and Chernetsov 2000; Bowlin and Wikelski 2008]). The results of measurements made in flushed birds that made short escape flights under unnatural conditions cannot be automatically generalised to migrants that were flying by sustained flight for hours. Both field and experimental data strongly suggest that the cost of migratory flight is considerably lower than is usually assumed. This is further confirmed by the data on wingbeat rate in several North American thrushes (Cochran et al. 2008). The wingbeat rate is closely correlated with flight power, and during nocturnal migratory flights it was on average 1.5 times lower than in the same individuals during their daytime foraging flights.

All this supports the earlier ideas of Dolnik (1969, 1971) that migratory flight is relatively cheap because migrants can change their aerodynamic quality when necessary. Our data provide further evidence supporting the suggestion that the current aerodynamic theory of avian flight needs revision (Videler 2005; Schmidt-Wellenburg et al. 2008).

References

Able KP (1982) Skylight polarization patterns at dusk influence migratory orientation in birds. Nature 299:550–551

Able KP (1989) Skylight polarization and the orientation of migratory birds. J Exp Biol 141:241–256

Able KP (1993) Orientation cues used by migratory birds: a review of cue-conflict experiments. Trends Ecol Evol 8:367–371

Agatsuma R, Ramenofsky M (2006) Migratory behaviour of captive white-crowned sparrows, *Zonotrichia leucophrys gambelii*, differs during autumn and spring migration. Behaviour 143:1219–1240

Åkesson S (1999) Do passerine migrants captured at an inland site perform temporary reverse migration in autumn? Ardea 87:129–137
Åkesson S, Hedenström A (2000) Wind selectivity of migratory flight departures in birds. Behav Ecol Sociobiol 47:140–144
Åkesson S, Alerstam T, Hedenström A (1996a) Flight initiation of nocturnal passerine migrants in relation to celestial orientation conditions at twilight. J Avian Biol 27:95–102
Åkesson S, Karlsson L, Walinder G, Alerstam T (1996b) Bimodal orientation and the occurrence of temporary reverse bird migration during autumn in south Scandinavia. Behav Ecol Sociobiol 38:293–302
Åkesson S, Walinder G, Karlsson L, Ehnbom S (2001) Reed warbler orientation: initiation of nocturnal migratory flights in relation to visibility of celestial cues at dusk. Anim Behav 61:181–189
Åkesson S, Walinder G, Karlsson L, Ehnbom S (2002) Nocturnal migratory flight initiation in reed warblers *Acrocephalus scirpaceus*: effect of wind on orientation and timing of migration. J Avian Biol 33:349–357
Alerstam T (1976) Nocturnal migration of thrushes (*Turdus* spp.) in southern Sweden. Oikos 27:457–475
Alerstam T (1990) Bird migration. Cambridge University Press, Cambridge
Berthold P (2001) Bird migration: a general survey, 2nd edn. Oxford University Press, Oxford
Biebach H (1985) Sahara stopover in migratory flycatchers: fat and food affect the time program. Experientia 41:695–697
Biebach H (1996) Energetics of winter and migratory fattening. In: Carey C (ed) Avian energetics and nutritional ecology. Chapman & Hall, NY
Blem CR (1990) Avian energy storage. In: Power DM (ed) Current ornithology, vol 7. Plenum Press, NY
Bolshakov CV (1981) Rekonstruktsiya polnoy kartiny nochnogo prolyota i effektivnost' obnaruzheniya ego raznymi metodami (Reconstruction of the total picture of nocturnal passage and effectiveness of several methods of its estimation). Proc Zool Inst 104:95–123
Bolshakov CV (1992) Vechernie peremescheniya i start nochnogo migratsionnogo poleta u drozda-ryabinnika (*Turdus pilaris*): predvaritelnye rezultaty (Evening movements and the start of the nocturnal migratory flight in the fieldfare (*Turdus pilaris*): preliminary results). Proc Zool Inst 247:18–42
Bolshakov CV, Bulyuk VN (1999) Time of nocturnal flight initiation (take-off activity) in the European Robin *Erithacus rubecula* during spring migration: direct observations between sunset and sunrise. Avian Ecol Behav 2:51–74
Bolshakov CV, Bulyuk VN (2001) New comprehensive systematic data concerning the time of nocturnal departure in some passerine migrants in autumn. Ring 23:131–137
Bolshakov CV, Chernetsov N (2004) Initiation of nocturnal flight in two species of long-distance migrants (*Ficedula hypoleuca* and *Acrocephalus schoenobaenus*) in spring: a telemetry study. Avian Ecol Behav 12:63–76
Bolshakov CV, Rezvyi SP (1998) Time of nocturnal flight initiation (take-off activity) in the European Robin *Erithacus rubecula* during spring migration: visual observations between sunset and darkness. Avian Ecol Behav 1:37–49
Bolshakov CV, Bulyuk VN, Sinelschikova A (2000) Study of nocturnal departures in small passerine migrants: retrapping of ringed birds in high mist-nets. Vogelwarte 40:250–258
Bolshakov CV, Shapoval AP, Zelenova NP (2001) Results of bird trapping and ringing by the Biological Station ''Rybachy'' on the Courish Spit: long-distance recoveries of birds ringed in 1956–1997. Part 1. Avian Ecol Behav (Suppl 1):1–126
Bolshakov CV, Shapoval AP, Zelenova NP (2002a) Results of bird trapping and ringing by the Biological Station ''Rybachy'' on the Courish Spit: controls of birds ringed outside the Courish Spit in 1956–1997. Part 1. Avian Ecol Behav (Suppl 5):1–106
Bolshakov CV, Švažas S, Žalakevičius M (2002b) Nocturnal migration of thrushes in the Eastern Baltic region. Akstis, Vilnius

Bolshakov C, Bulyuk V, Chernetsov N (2003a) Spring nocturnal migration of Reed Warblers *Acrocephalus scirpaceus*: departure, landing and body condition. Ibis 145:106–112

Bolshakov CV, Bulyuk VN, Mukhin A, Chernetsov N (2003b) Body mass and fat reserves of Sedge Warblers during vernal nocturnal migration: departure versus arrival. J Field Ornithol 74:81–89

Bolshakov CV, Chernetsov N, Mukhin A, Bulyuk VN, Kosarev V, Ktitorov P, Leoke D, Tsvey A (2007) Time of nocturnal departures in European robins, *Erithacus rubecula*, in relation to celestial cues, season, stopover duration and fat stores. Anim Behav 74:855–865

Bowlin MS, Wikelski M (2008) Pointed wings, low wing loading and clam air reduce migratory flight costs in songbirds. PLoS ONE 3:e2154

Bulyuk VN (2006) At what time of the day do passerine nocturnal migrants arrive at their breeding sites? Ardea 94:132–139

Bulyuk V, Chernetsov N (2000) Two migratory flights of sedge warblers *Acrocephalus schoenobaenus* from Finland to Estonia. Ornis Svecica 10:79–83

Bulyuk VN, Mukhin A, Fedorov VA, Tsvey A, Kishkinev D (2000) Juvenile dispersal in Reed Warblers *Acrocephalus scirpaceus* at night. Avian Ecol Behav 5:45–63

Bulyuk VN, Tsvey A (2006) Timing of nocturnal autumn migratory departures in juvenile European robins (*Erithacus rubecula*) and endogenous and external factors. J Ornithol 147:298–309

Casement MB (1966) Migration across the Mediterranean observed by radar. Ibis 108:461–491

Chernetsov NS (1999) Migratsionnye strategii kamyshevok *Acrocephalus* spp. v predelakh Evropy (Migratory strategies of *Acrocephalus* warblers within Europe). Dissertation, Zoological Institute RAS

Chernetsov N (2006) Habitat selection by nocturnal passerine migrants en route: mechanisms and results. J Ornithol 147:185–191

Chernetsov N, Kishkinev D, Mouritsen H (2008) A long-distance avian migrant compensates for longitudinal displacement during spring migration. Curr Biol 18:188–190

Cochran WW (1987) Orientation and other migratory behaviours of a Swainson's thrush followed for 1,500 km. Anim Behav 35:927–929

Cochran WW, Montgomery GG, Graber RR (1967) Migratory flights of *Hylocichla* thrushes in spring: a radiotelemetry study. Living Bird 6:213–225

Cochran WW, Mouritsen H, Wikelski M (2004) Migrating songbirds recalibrate their magnetic compass daily from twilight cues. Science 304:405–408

Cochran WW, Bowlin MS, Wikelski M (2008) Wingbeat frequency and flap-pause ratio during natural migratory flight in thrushes. Integr Comp Biol 48:143–151

Delingat J, Bairlein F, Hedenström A (2008) Obligatory barrier crossing and adaptive fuel management in migratory birds: the case of the Atlantic crossing in northern wheatears (*Oenanthe oenanthe*). Behav Ecol Sociobiol 62:1069–1078

Dolnik VR (1969) Bioenergetika letyaschey ptitsy (Bioenergetics of a flying bird). Zh Obschei Biologii 30:273–291

Dolnik VR (1971) Energetika pereletov ptits (Eneregtics of avian migration). In: Poznanin LP (ed) Voprosy ornitologii (Topics of ornithology), vol 2, pp 52–81

Dolnik VR (1995) Resursy energii i vremeni u ptits v prirode (Energy and time resources in free-living birds). Nauka, St Petersburg

Dolnik VR, Gavrilov VM (1973) Energy metabolism during flight of some passerines. In: Bykhovskii BE (ed) Bird migrations, ecological and physiological factors. Halstead Press, NY

Drury WH, Nisbet ICT (1964) Radar studies of orientation of songbird migrants in southeastern New England. Bird-Banding 35:69–119

Emlen ST (1975) Migration: orientation and navigation. In: Farner DS, King JR, Parkes KC (eds) Avian Biology, vol 5. Academic Press, NY

Emlen ST (1980) Decision making by nocturnal bird migrants: the integration of multiple cues. In: Nohring R (ed) Acta XVII International Ornithological Congress. Deutsche Ornithologen-Gesellschaft, Berlin

Gauthreaux SA Jr (1971) A radar and direct visual study of passerine spring migration in southern Louisiana. Auk 88:343–365
Gavrilov VM (2011) Energy expenditures for flight, aerodynamic quality, and colonization of forest habitats by birds. Biol Bull 38:779–788
Hall S (1996) Time timing of post-juvenile moult and fuel deposition in relation to the onset of autumn migration in reed warblers *Acrocephalus scirpaceus* and sedge warblers *Acrocephalus schoenobaenus*. Ornis Svecica 6:89–96
Holmes RT, Sawyer RH (1975) Oxygen consumption in relation to ambient temperature in five species of forest-dwelling thrushes (*Hylocichla* and *Catharus*). Comp Biochem Physiol A 50:527–531
Karlsson H, Nilsson C, Bäckman J, Alerstam T (2011) Nocturnal passerine migration without tailwind assistance. Ibis 153:485–493
Kerlinger P, Moore FR (1989) Atmospheric structure and avian migration. In: Johnston RF (ed) Current ornithology, vol 6. Plenum Press, NY
Klaassen M, Kvist A, Lindström Å (2000) Flight costs and fuel composition of a bird migrating in a wind tunnel. Condor 102:444–451
Komenda-Zehnder S, Liechti F, Bruderer B (2002) Is reverse migration a common feature of nocturnal bird migration?—an analysis of radar data from Israel. Ardea 90:325–334
Lindström Å, Klaassen M, Kvist A (1999) Variation in energy intake and basal metabolic rate of a bird migrating in a wind tunnel. Funct Ecol 13:352–359
Martin G (1990) Birds by night. A&C Black, London
Masman D, Klaassen M (1987) Energy expenditure during free flight in trained and free-living Eurasian kestrels (*Falco tinnunculus*). Auk 104:603–616
Mills AM, Thurber BJ, Mackenzie SA, Taylor PD (2011) Passerines use nocturnal flight for landscape-scale movements during migratory stopover. Condor 113:597–607
Moore FR (1987) Sunset and the orientation behaviour of migrating birds. Biol Rev 62:65–86
Moore FR, Aborn DA (1996) Time of departure by summer tanagers (*Piranga rubra*) from a stopover site following spring trans-Gulf migration. Auk 113:949–952
Muheim R, Phillips JB, Åkesson S (2006) Polarized light cues underlie compass calibration in migratory songbirds. Science 313:837–839
Muheim R, Åkesson S, Phillips JB (2007) Magnetic compass of migratory Savannah sparrows is calibrated by skylight polarization at sunrise and sunset. J Ornithol 148(Suppl 2):S485–S494
Mukhin A (2004) Night movements of young reed warblers (*Acrocephalus scirpaceus*) in summer: Is it postfledging dispersal? Auk 121:203–209
Mukhin A, Kosarev V, Ktitorov P (2005) Nocturnal life of young songbirds well before migration. Proc R Soc Lond B 272:1535–1539
Mukhin A, Grinkevich V, Helm B (2009) Under cover of darkness: nocturnal life of diurnal birds. J Biol Rhythms 24:225–231
Newton I (2008) The migration ecology of birds. Academic Press, London
Palmgren P (1949) On the diurnal rhythm of activity and rest in birds. Ibis 91:561–576
Parslow JLF (1969) The migration of passerine night migrants across the English Channel studied by radar. Ibis 111:48–79
Pennycuick CJ, Alerstam T, Hedenström A (1997) A new low-turbulence wind tunnel for bird flight experiments at Lund University, Sweden. J Exp Biol 200:1441–1449
Ramenofsky M, Agatsuma R, Barga M, Cameron R, Harm J, Landys M, Ramfar T (2003) Migratory behavior: new insights from captive studies. In: Berthold P, Gwinner E, Sonnenschein E (eds) Avian migration. Springer, Berlin
Ramenofsky M, Agatsuma R, Ramfar T (2008) Environmental conditions affect the behavior of captive, migratory white-crowned sparrows. Condor 110:658–671
Rayner JMV (1994) Aerodynamic corrections for the flight of bats and birds in wind tunnels. J Zool 234:537–563
Richardson W (1978) Timing and amount of bird migration in relation to weather: a review. Oikos 30:224–272

Richardson W (1990) Timing of bird migration in relation to weather: updated review. In: Gwinner E (ed) Bird migration. Springer, Berlin
Salewski V, Herremans M, Liechi F (2010) Migrating passerines can lose more body mass reversibly than previously thought. Ring Migr 25:22–28
Sandberg R (1991) Sunset orientation of robins, *Erithacus rubecula*, with different fields of sky vision. Behav Ecol Sociobiol 28:77–83
Schmaljohann H, Naef-Daenzer B (2011) Body condition and wind support initiate the shift of migratory direction and timing of nocturnal departure in a songbird. J Anim Ecol 80:1115–1122
Schmaljohann H, Liechti F, Bruderer B (2007a) Songbird migration across the Sahara: the non-stop hypothesis rejected! Proc Royal Soc Lond B 274:735–739
Schmaljohann H, Liechti F, Bruderer B (2007b) An addendum to 'songbird migration across the Sahara: the non-stop hypothesis rejected!'. Proc Royal Soc Lond B 274:1919–1920
Schmaljohann H, Becker PJJ, Karaardic H, Liechti F, Neaf-Daenzer B, Grande C (2011) Nocturnal exploratory flights, departure time, and direction in a migratory songbird. J Ornithol 152:439–452
Schmidt-Wellenburg CA, Engel S, Visser GH (2008) Energy expenditure during flight in relation to body mass: effects of natural increases in mass and artificial load in rose coloured starlings. J Comp Physiol B 178:767–777
Seewagen CL (2008) An evaluation of condition indices and predictive models for nonivasive estimates of lipid mass of migrating common yellowthroats, ovenbirds, and Swainson's thrushes. J Field Ornithol 79:80–86
Shapoval AP (1981) Vidovoy i kolichestvenny sostav migriruyuschikh ptits, popadayuschikh nochyu v statsionarnye lovushki na Kurshskoy kose (Specific and quantitative composition of the birds captured at night in stationary traps on the Courish Spit). In: Michelson HA (ed) Abstracts 10th Baltic ornithological conference, Part 1. Riga, Zīnatne
Švažas S (1993) Weather factors affecting migratory take-off of nocturnal migrants in autumn. Acta Ornithol Lithuanica 7–8:27–35
Taylor PD, Mackenzie SA, Thurber BG, Calvert AM, Mills AM, McGuire LP, Guglielmo CG (2011) Landscape movements of migratory birds and bats reveal and expanded scale of stopover. PLoS ONE 6:e27054
Videler JJ (2005) Avian flight. Oxford University Press, Oxford
Wikelski M, Tarlow EM, Raim A, Diehl RH, Larkin RP, Visser GH (2003) Costs of migration in free-flying songbirds. Nature 423:704
Yablonkevich ML (1987) Massa tela, vodnye i energeticheskie komponenty ptits v periody vesennego i osennego proletov na o. Barakelmes Aralskogo morya (Body weight, water and energy components of birds migrating in spring and autumn across the Aral Sea Island, Barsakelmes). Proc. Zool. Inst. 173:49–71
Zehnder S, Åkesson S, Liechti F, Bruderer B (2002) Observation of free-flying nocturnal migrants at Falsterbo: occurrence of reverse flight directions in autumn. Avian Sci. 2:103–113
Zimin VB (2003) Body mass variability in juvenile Robins *Erithacus rubecula* in the Ladoga area. Avian Ecol Behav 10:1–31

Chapter 8
Migratory Flights and Stopovers: Organisation of Migration

Abstract In this chapter I summarise the contents of this monograph and propose the model that four main groups of factors influence the decision to take off: current fuel stores, fuel deposition rate at a given stopover site, weather factors (mainly wind assistance) and geographic position relative to the optimal rate of migration. I also discuss the principles of organisation of migration in songbirds and different selection pressures that may govern migration from the breeding to the non-breeding area and back. This chapter also contains the main conclusions, and after them I outline the future research directions that may bring further progress to the field. The alternation of flights and stopovers is apparently organised in some manner. The aim of our study is to identify the principles that govern this organisation of migration. Flights are dependent on stopovers in an apparent way: during flights, the energy stored during the stopovers is used. Therefore it is necessary to obtain estimates of the energy deposition rate at stopovers and of the rate of using energy during flights.

8.1 The Importance of Fuel Deposition Rate

As shown in Chap. 3, FDR in the wild is usually rather low, and the maximum known values follow the equation $FDR_{max} = 2.17\ mass^{-0.34}$, where mass is body mass in kilograms (Lindström 2003). This means that a Eurasian reed warbler with the lean body mass of ca. 10.0 g under optimal conditions can increase its body mass by 1.04 g per day. Assuming the energy density of fuel stores of 21.6 kJ g^{-1} (Klaassen et al. 2000) and flight power of 5.80 kJ h^{-1} (Sect. 7.7), one day of fuelling at the maximum possible rate provides a Eurasian reed warbler with energy sufficient for flying during 3.9 h. The duration of an average nocturnal migratory flight during spring passage in this species is 4.1 h (Bolshakov et al. 2003a), i.e. the energy gained during one day of feeding ad libitum is nearly sufficient for

N. Chernetsov, *Passerine Migration*, DOI: 10.1007/978-3-642-29020-6_8,

one night of flight. For comparison, a European robin can gain a maximum of 29.2 kJ of energy in one day, with allows it to fly during 3.7 h, assuming the flight cost of 7.1 BMR and BMR = 26.45 kJ day^{-1}, which is the mean of the four published estimates: 26.0 (Gavrilov 1980); 29.1 (Gavrilov 1985); 24.3 and 26.4 kJ day^{-1} (Gavrilov 1981).

These calculations predict the maximum possible fuel deposition rate. The FDR that is reached in the wild is usually much lower due to food shortage, competition, predation risk etc. (Chap. 3). Besides, in the first 1–2 days upon arrival at stopover FDR may be very low, or even negative (Szulc-Olech 1965; Pettersson 1983; Hansson and Pettersson 1989; Mädlow 1997; Titov 1999; Sect. 3.5). Obviously, such a hitch at the beginning makes the mean FDR across the overall time of stopover much lower. Therefore, in the real world migrants need much longer than a single foraging day to gain energy necessary to fly during 3.7–3.9 h.

It should be also kept in mind that the FDR values observed in the wild always show a large dispersion, and the significant predictors (e.g. initial fuel load) often leave a great deal of variation unexplained (Chernetsov and Titov 2001; Chernetsov et al. 2007). As an example, we can look into FDR values of Eurasian reed warblers, sedge warblers and garden warblers at several European and North African sites during autumn migration (Schaub and Jenni 2000). In the Eurasian reed warbler the proportion of explained variance in FDR in the best generalised linear model varied between 21.2 – 64.9% (13 sites); in the sedge warbler, between 26.0 and 38.4% (seven sites, the value for Mettnau, SW Germany [97.4%] is omitted because it is based on the analysis of only 10 recaptures); in the garden warbler, between 0.1 and 69.2% (12 sites). The highest proportion of explained variance among the sites where >100 recaptures were included was 42.5% (FDR of Eurasian reed warblers in Mettnau). The predictors were stopover duration, initial body mass, the progress of season and moult status. Thus, about one-half of variance in FDR usually remains unexplained.

8.2 Factors that Govern Departure Decisions

The main currency that shapes the organisation of migration is the energy necessary for flight (and safety, i.e. the probability to survive; see Chap. 4). As shown by both field and experimental data, fat individuals are more prone to take off than lean ones (Wang and Moore 1993; Jenni and Schaub 2003; Fusani et al. 2009). However, it is not possible to construct a realistic and generally a heuristically useful model of the migratory strategy of passerines on the basis of energetic considerations alone. As shown in Chap. 4, the attempts made in the framework of the optimal migration theory cannot be called successful. They did not result in the models that would fit the data well enough, because apart from the energetic (i.e. endogenous) factor, the behaviour of the migrants is strongly modified by the environment.

Apart from the variation in fuel deposition rate during the stopover period, departure decisions are strongly influenced by the weather, mainly by wind speed and direction (Schaub et al. 2004; Tsvey et al. 2007; Arizaga et al. 2011; Schmaljohann et al. 2011; Schmaljohann and Naef-Daenzer 2011). It should be mentioned that Victor Bulyuk and Arseny Tsvey (*in litt.* 2011) analysed weather conditions on the nights of departures of European robins that stopped over on the Courish Spit for more than one day and were captured in mist-nets at take-off. The weather on the nights of departure of such birds was not significantly different from the that in the preceding nights. However, it should be kept in mind that some of the recorded departures could be for exploratory flights (Schmaljohann et al. 2011; Mills et al. 2011), or the birds might abort their departures. This may or may not be recorded when radio-tracking the birds, but captures in elevated mist-nets do not allow to distinguish between the 'true' (or 'successful') migratory departures and the exploratory or aborted flights.

What happens after an aborted flight? In practice, the stopover continues. Does the migrant continue to gain fuel if the environmental conditions permit it? Such situations may occur not infrequently in September and October in northern and central Europe, when the weather conditions unfavourable for flight (opposing westerly and south-westerly winds) are favourable for refuelling (relatively high air temperature; Chernetsov 2002). Isolated cases of captures of very heavy birds not in front of a barrier, i.e. of European robins with the body mass exceeding 20 g (i.e. fuel load ca. 50%), sedge warblers weighing 18–19 g (fuel load ca. 80–90%), a great reed warbler weighing 45.5 g (Koleček 2006) seem to support this viewpoint. It should be however kept in mind that such captures of very fat individuals are very rare. On the other hand, as we have already mentioned more than once, mist-netting probability of a songbird depends on its mobility, and the mobility of the individuals with high fuel loads may be very low. Because of that, heavy migrants may be strongly underrepresented in captures and be more common in the wild than usually assumed (Kosarev and Kobylkov 2010).

The weather, including wind, may influence not only migratory departures but also the decision to cease flight. When crossing the Sahara, passerine nocturnal migrants continue their flights after daybreak if they enjoy following winds and land in the desert if the wind is opposing (Schmaljohann et al. 2007a). Emergency landing at night is well known when flight conditions quickly deteriorate. For instance, mass emergency landing of migrating thrushes, mainly song thrushes, was recorded in Lithuania on 23 October 1990 when the weather favourable for migratory flight was sharply replaced by low clouds and fog (Bolshakov et al. 2002).

The most realistic qualitative model of behaviour of migrants at stopover may be presented in the following way. Any migrant is influenced by at least four groups of factors that are to some extent independent on each other and thus may be represented as orthogonal axes in the four-dimensional hyperspace: (1) current energy status; (2) habitat quality that makes it possible to reach a certain FDR; (3) weather conditions, first of all wind assistance; and (4) position on the migratory route with respect to the individual schedule of migration. Most prone to continue migration, i.e. to depart, are fat birds at a poor site under weather conditions favourable for flight that are delayed on the migratory route (i.e. because of unfavourable weather in the preceding days). The position of a migrant along any of these axes influences its readiness to migrate: under favourable weather conditions both fat birds depart (that would have departed also in poorer weather) and their leaner conspecifics that in worse weather would have remained at stopover. There is evidence that the optimal schedule of migration may be individually adjusted, as shown by the data at the onset of spring migration of bar-tailed godwits from New Zealand (Battley 2006; Conklin et al. 2010) and from Portugal (Lourenço et al. 2011). If migration is much delayed (like in late May—early June 1974 on the southern coast of the Gulf of Finland in the European robin; Bolshakov and Rezvyi 1998), nearly all individuals depart for migratory flights, even those whose fuel stores are low. This is probably what Tatiana Blyumental and Victor Dolnik meant when they wrote about 'carrying along of the lean individuals by the fat ones' when explaining how the waves of migration were formed (Blyumental et al. 1967; Dolnik and Blyumental 1967; Dolnik 1975). This 'carrying along' was postulated when discussing the migration of fringillid finches, i.e. short- and medium-distance diurnal migrants (Dolnik 1975); its existence in solitary nocturnal migrants seems dubious. After the periods of strong opposing winds which are very unfavourable for migration on the Courish Spit significantly more European robins take off under moderate opposing winds, i.e. with negative wind assistance (Bulyuk and Tsvey *in litt.* 2011). Recent radar studies have shown that songbirds perform migratory flights with negative wind assistance more often than hitherto assumed (Karlsson et al. 2011).

The influence of stopover habitat quality on the departure decision has been shown experimentally. European robins and pied flycatchers in migratory disposition stopped to show migratory restlessness (Zugunruhe) if after a period of fasting (which imitated migratory flight) they were fed ad libitum. Zugunruhe resumed when food access was again restricted, imitating poor stopover site (Merkel 1938, 1958; Biebach 1985; Gwinner et al. 1988). In the field experiment Eurasian reed warblers tape-lured into a suboptimal habitat (sand dunes on the Courish Spit) departed from the stopover site by nocturnal migratory flights (Ktitorov et al. 2010). All individuals, even the lean ones, left the area by nocturnal flights on the first night after arrival ($n = 10$).

Our model assumes that the values of each factor which releases flight are not fixed, but vary within certain limits which are defined by other factors. This makes it possible to explain why birds captured at migratory take-off show a broad variation of body mass and fuel load values (Sect. 7.2). This range of variation does not support the idea of the threshold fuel load releasing flights. The weather conditions under which migrants depart (Bulyuk and Tsvey 2006; Bolshakov et al. 2007; Tsvey et al. 2007; Bulyuk and Mukhin 2010) and fly (Richardson 1978, 1990; Bolshakov 1981; Karlsson et al. 2011) also vary broadly. Some individuals start and perform migratory flights under the most inclement weather conditions, which permitted Victor Dolnik to claim that "with respect to birds it is a mistake to use the words 'flying' or 'non-flying' weather: migratory birds are practically all-weather aircrafts" (Dolnik 1975, p. 41). Certainly this does not mean that the weather conditions do not influence bird migration. However, whatever factor we analyse in isolation, the migratory flights may be performed (and start) at so broad a range of its values that a migratory flight may be initiated by any individual in any condition. Only the analysis of the combination of the aforementioned factors may help identify the patterns.

It should be mentioned that this model is actually just an application of the well-known limiting factor rule (Begon et al. 2006) to movement ecology of birds: of the four suggested groups of factors of the greatest importance for the departure decisions is the one which is the closest to the critical (i.e. prohibiting) value.

It should be also kept in mind that the independence of the four groups of factors (and, as a consequence, the orthogonality of the representing axes) is rather conventional. The current energy condition of the migrant is only independent of the habitat quality at the stopover site immediately after arrival (and even habitat selection during arrival may be and most probably is condition-dependent). During stopover the latter factor greatly influences the former one. Current weather (its suitability for migratory flights) is often correlated with the weather during the preceding days, which may influence the fuel deposition rate (especially in early arriving individuals in spring) and the position of the bird with respect to its migratory schedule. However, the assertion that 'every factor depends on all the others' may be formally correct but is not very enlightening. Therefore I suggest that there is a point in isolating groups of relatively independent factors but the conditional character of their independence should be retained.

8.3 Series of Migratory Flights and Waves of Passage

Small passerines that resume migration after a successful prolonged stopover usually appear to make several nocturnal flights in succession. A series of migratory flights starts when migrants accumulate sufficient fuel stores; its onset seems to be mainly governed by energy and the spatiotemporal migratory programme (Dolnik and Blyumental 1967; Dolnik 1968, 1975; Berthold 1996, 2001; Jenni and Schaub 2003). It seems that migrants that resume migration after prolonged stopovers may depart under a broad range of weather conditions (Bulyuk and Tsvey 2006; Tsvey et al. 2007; Bulyuk and Mukhin 2010). If the weather is favourable for migratory flights, migration may continue beyond the first night. A relatively small energy cost of flying in good weather (low air turbulence, significant wind assistance) further favours flying during several nights. Between these nocturnal flights migrants make one-day stopovers, i.e. they are transients at the respective stopover sites. This tactics of migration seems to be typical of songbird long-distance migrants both in spring (Bolshakov et al. 2003a, b; Tøttrup et al. 2012) and in autumn, as shown by the analysis of ring recoveries (Fransson 1995; Hall-Karlsson and Fransson 2008).

The study of the temporal schedule of nocturnal departures of European robins showed that their readiness to take off after one-day stopovers and partly departure time were mainly defined by the weather in the preceding night, i.e. during the previous flight (Bulyuk and Tsvey 2006). European robins can continue the series of migratory flights if they have previously migrated with following winds and have not depleted their fuel stores. Conversely, if the flight occurred under unfavourable winds and resulted in large expenditure of energy, the series of flights is terminated (Dolnik and Blyumental 1967; Dolnik 1975). It cannot also be ruled out that progress towards the goal of migration may also influence the decision to continue flights. It is generally assumed that first-time migrants have no information about the goal of their migration (Gwinner and Wiltschko 1978; Berthold 1990, 1996). However, evidence is accumulating that first-autumn migrants can control their position on the migratory route on the basis of external references, in particular of the geomagnetic field parameters (Beck and Wiltschko 1988; Wiltschko and Wiltschko 1992; Fransson et al. 2001; Kullberg et al. 2007; Chernetsov et al. 2008; Henshaw et al. 2008, 2009).

Anyway, wind assistance is apparently one of the most important extrinsic factors that govern the decision to continue migration or terminate it. It should be emphasized that before migrants take off, they seem to have no 'miraculous' method to know the high-altitude wind. The only way to know the wind at the normal flight altitudes is to take off and to test it, as suggested by exploratory flights of northern wheatears on Helgoland (Schmaljohann et al. 2011) and by aborted departures of European robins on the Courish Spit. Changes in wind conditions may synchronise the cycles of migratory activity of individual migrants (Dolnik and Blyumental 1967; Dolnik 1975; Schaub et al. 2004). As a result, the dynamics of passage often (but not always) has a pronounced wave-like pattern, when peaks of passage are alternated with much quieter nights (Bolshakov 1981; Erni et al. 2002). As some migrants may be at different stages of the series of migratory flights (some may be starting, some continuing, and some finishing them) individual migrants that arrive at a stopover site on the same night, may show broad variation in fuel stores and stopover duration (Tsvey et al. 2007).

A series of migratory flights comes to an end when fuel stores of migrants are depleted and/or when the weather deteriorates. Using such migratory tactics during autumn migration may be adaptive to the unpredictable and often unstable weather. In autumn in northern and north-eastern Europe such behaviour allows the birds to migrate towards the south-west in spite of dominating opposing winds and allow them to use improvements in the weather conditions in the optimal way. The same tactics is optimal during spring migration in Europe, when warm weather which significantly improves feeding conditions for insectivorous migrants is usually caused by intrusions of warm air from the Mediterranean, which also provide favourable conditions for northbound flights (Chernetsov 2002; Bolshakov et al. 2003a). In other regions and seasons, when opposing winds are less frequent, the ability to fine-tune migratory behaviour to the weather conditions might be less crucial.

The results of such tactics of migration are apparent from the analysis of daily variation in trapping numbers. New individuals arrive at stopover after nights with any wind direction. However, the numbers of European robins peak after the nights with following winds (Bulyuk and Tsvey *in litt.* 2011). On some nights the numbers of arriving migrants also increased in the nights with weak or moderate opposing winds, when they followed the nights with strong opposing winds (Bolshakov and Rezvyi 1998; Erni et al. 2002). Therefore, the weather conditions may synchronise the series of migratory flights performed by different individuals and help formation of the waves of passage recorded by visual observations or in trapping projects (Blyumental et al. 1967; Dolnik 1975).

Songbird migrants take off under different weather conditions, but quickly abort their flights when the wind is unfavourable (Schmaljohann et al. 2011). If the wind is favourable (in spring) or at least less unfavourable than in the preceding days (in autumn), a large number of birds continue flight, so that a migratory wave is formed. If the favourable weather persists, the fuel stores of migrants are used up rather slowly, so that migratory flights can be performed during several nights is succession. In such case a strong wave of passage is observed, like it was shown for fieldfares during spring migration (Bolshakov 1992). In the areas where the weather conditions during the migratory season are very stable (e.g. in the Central Asian deserts), the flow of migrants may be very uniform during several weeks, without any waves.

8.4 Spring Versus Autumn Migration

Spring and autumn migration (to be more exact, migration to and from the breeding quarters) mainly differ in that in spring most adult passerines return to their previous breeding area (i.e. show breeding site fidelity), and many yearlings head for the area that they have imprinted as the future breeding site during postfledging movements in the previous year (Sokolov 1997; Newton 2008; Grinkevich et al. 2009). Without discussing the mean distance of natal dispersal in passerines (e.g. Paradis et al. 1998), we can safely claim that a very significant proportion of first-time breeders returns to the area whose linear size is by 2–3 orders of magnitude smaller than their migratory distance. This has been repeatedly shown by ringing recoveries. It means that in spring, all or most migrants have a certain migratory goal, whereas during autumn migration, only adult experienced individuals may have a goal. First-autumn migrants fly towards the areas they have never visited before. The currently accepted clock-and-compass concept assumes that juvenile migrants have no inherited knowledge of their migratory destination except of the (necessarily general) inherited programme but the recent data suggests that this view may be challenged (see Sect. 1.2).

It cannot be ruled out that the existence of a certain narrow migratory target in spring and its absence in many first-time migrants in autumn may influence their migratory behaviour. It has been shown that at least some individuals of long-distance songbird migrants arrive in spring at their breeding sites by nocturnal flights (Bulyuk 2006) and do not perform any slow search in the daytime as it has been believed for a long time (Heinroth and Heinroth 1941). These data indicate a very precise, pinpoint navigation of nocturnal migrants during their flights, with an accuracy of ca. 1 km. It is most likely that these birds took off from different last stopover areas and started their flights at different time after sunset. The distance to the goal of migration may be an important factor that influences the timing of migratory departure in spring (Bolshakov and Bulyuk 1999; Bolshakov et al. 2007).

Apart from these factors that are universal, variation caused by regional weather features, for instance, thc direction of the prevailing winds may exist. In central and northern Europe in spring warm periods, when the activity of invertebrates increases and thus food availability for insectivorous migrants improves (Chernetsov and Manukyan 1999, 2000), are usually caused by the intrusions of warm air from the Mediterranean region. These southerly or south-westerly winds provide good wind assistance for most songbirds migrating in spring in this region. Therefore, in spring the same synoptic weather situations are favourable for both migratory flights and stopovers. It allows the migrants to perform several flights in succession and to move towards their migratory target quickly (Bolshakov et al. 2003a, b). This behaviour is very adaptive because in spring many avian migrants, especially males, benefit from early arrival at their breeding grounds, and the bonus for early arrival may be very significant at least for some individuals (Kokko 1999; Forstmeier 2002).

In autumn, in the same region warm south-westerly winds influence favourably food availability for insectivors, but are opposing to and thus unfavourable for migratory flights. As a result, wind selectivity of passerine migrants may vary between different years: is some years migrants are less selective than in others, probably because of the rarer occurrence of favourable winds (Tsvey et al. 2007). This situation is a special case of the more general rule described in Sect. 8.2.

Apparently, in other regions of the Earth weather (mainly wind) conditions prevailing in the respective seasons may influence bird migration in a different manner. Europe and eastern North America are relatively well studied in this respect, whereas the patterns that occur in other regions are very poorly known and cry for research into the regional aspects of avian migration (e.g. studies in Central Asia: Dolnik 1990; Bolshakov 2002, 2003; Raess 2008; in temperate East Asia: Wang et al. 2006; Yamaguchi et al. 2008, 2012). These studies can also shed light on the basic patterns of avian migration.

8.5 Annual Movements of a Typical Long-Distance Passerine Nocturnal Migrant

Let us imagine a typical long-distance avian migrant, for example a Eurasian reed warbler, breeding in Eastern Europe. Juvenile Eurasian reed warblers start their nocturnal postfledging movements early, when 35–40 days old (Mukhin et al. 2005).

The function of these nocturnal flights is still unclear. They might be necessary to form the navigational target to which they will try to return the next spring, or to develop flying and orientation abilities (Mukhin et al. 2005). However, the birds do not start to migrate, i.e. to consequently move in the migratory direction until they are 50–55 days old (Mukhin 2004). By the age of 60 days all juveniles will have left their natal area and started autumn migration (Chernetsov and Mukhin 2001). Therefore, most Eurasian reed warblers start autumn migration in mid—late August. Most adults start their autumn migration even earlier, in late July, and overlap it with body moult, even though this overlap is more typical of passerines breeding further north, in the boreal forest and tundra (Panov 2011).

It has been shown in several long-distance passerine migrants (the most considerable material was obtained for the Eurasian reed warbler) that adults start their first nocturnal flight during autumn migration with relatively large fuel stores, on average 17.6% of their lean body mass (SD = 5.6, n = 6; Bulyuk 2010). The mean body mass of adult Eurasian reed warblers that took off for their first migratory flight from the breeding area was practically indistinguishable from the values typical of adult transients (Bulyuk 2010). It strongly suggests that adults start migration in the developed migratory disposition, and the first nocturnal flight may be rather long-range. The preliminary radio-tracking data suggests that the situation in first-autumn Eurasian reed warblers may be similar (Kosarev and Kobylkov 2010). However, in many species and populations of long-distance migrants migratory speed increases with the progress of migration (Hedenström and Pettersson 1987; Ellegren 1990, 1993), therefore, early during autumn migration flights may be shorter, and stopovers longer (Panov 2012).

Eurasian reed warblers from the Eastern Baltic migrate through the Iberian Peninsula (Chernetsov 1999), the distance to which is ca. 2,000 km; other long-distance migrants cover some 1,500 km to the Mediterranean coast. The mean speed of migration (i.e. movement along the migratory route with stopover time considered) in Eurasian reed warblers ringed in central Sweden was on average 39 km day^{-1} (Bensch and Nielsen 1999), i.e. the birds should need some 50 days to cover 2,000 km. It is however conceivable that the speed was underestimated in this study, because some individuals had been ringed during postfledging movements before their actual migratory departure. In other long-distance songbird migrants the estimates of migratory speed are higher: in Swedish sedge warblers it was on average 55 km day^{-1} (Bensch and Nielsen 1999), in *Sylvia* warblers varied between 43 and 93 km day^{-1} depending on species and the area of origin (Fransson 1995). Assuming such estimates, the Mediterranean may be reached more quickly, in 25–40 days. Red-backed shrikes supplied with geolocators travelled from the southern Baltic coast to SE Europe with the average speed of 101 km day^{-1} and reached that area in only 9 days (SD = 5.1; n = 9; Tøttrup et al. 2012).

In late September our Baltic Eurasian reed warbler reaches the northern edge of the extensive ecological barrier which is formed by the Mediterranean and the Sahara, and starts to fuel up. Until now, migration occurred without accumulating large fuel loads that would significantly increase the energy cost of flying

(Sect. 3.5), but now the migrant has to accumulate significant fuel stores (at least 50% and up to 100% of its lean body mass) that make the flight significantly more expensive, in line with the predictions of the aerodynamic theory.

The current data obtained by large-scale radar studies of passerine migration in the Sahara clearly show that the main strategy of crossing the desert is flight during the day and rest in the desert during the night (Schmaljohann et al. 2007a, b). The concept of non-stop flight across the desert suggested by Reginald E. Moreau half a century ago suggesting that migrants fly non-stop during several days (Moreau 1961) has raised doubts for quite some time (Bairlein 1985, 1988; Biebach et al. 1986), and now can be considered refuted (Schmaljohann et al. 2007a, b; Salewski et al. 2010). However, from the energetic viewpoint it makes the situation for the migrants more, not less, grave: during the diurnal rest the distance to the goal (relatively suitable stopovers areas on the northern edge of the Sahel) is not reduced, and the energy continues to be consumed (assumed at ca. 0.5% of body mass per hour; Meijer et al. 1994; Salewski et al. 2010).

After crossing the Sahara in tropical Africa different passerine migrants utilise very different spatial strategies. Some species, like the pied flycatcher (Salewski et al. 2002) or bluethroat (Markovets and Yosef 2005) occupy a territory where they spend the whole winter and where they recur every year. Other long-distance migrants, e.g. the willow warbler or garden warbler, move broadly within Africa, so that their movements may be called intra-African migration (Jones 1995; Salewski et al. 2002; Ottosson et al. 2005). Red-backed shrikes remain for 1–2 months in the Sahel/savannah zone of southern Sudan before they proceed to their final winter quarters in Botswana/Angola (Tøttrup et al. 2012). I suggest that their stay in Sudan, which is also typical of several other Palaearctic migrants, namely the marsh warbler, great reed warbler and whitethroat (Yohannes et al. 2009b), should be called the first winter quarter rather than stopover (cf. Introduction), and their movements between the first and the final winter quarters is intra-African migration. Little is known about the physiological basis of these movements (Terrill 1990).

The onset of spring migration is very poorly studied. It is one of the least known periods of the annual cycle in long-distance passerine migrants; recently, when radio-tracking has greatly advanced our knowledge of the postfledging period before the onset of autumn migration (Vega Rivera et al. 1998, 2003; Mukhin 2004; Mukhin et al. 2005), it probably became the least studied period.

It is believed that the onset of spring migration is under endogenous control (Berthold 1996), but recently some data has became available suggesting that this trait, too, may be modified by the environment, in particular, by precipitation in the wintering area (Sokolov and Kosarev 2003). In any case, migratory speed in spring is much higher than in autumn, as shown by the analysis of ring recoveries (Fransson 1995; Hall-Karlsson and Fransson 2008; Yohannes et al. 2009a) and by geolocator tracking data (Stutchbury et al. 2009; Heckscher et al. 2011; Tøttrup et al. 2012). It is usually assumed that the main reason for that is the necessity to arrive first to the breeding areas, because first arrivals, especially males, can occupy optimal breeding territories (Kokko 1999). As mentioned earlier, in spring

European long-distance migrants enjoy the weather situations favourable for migratory flights simultaneously with the weather situations favourable for foraging and fuelling. In the very recent years, information on the departure time and travel speed of songbirds from their winter quarters in South America (Stutchbury et al. 2009) and Africa (Tøttrup et al. 2012). Spring migration may be remarkably rapid: two purple martins travelled from Brazil to Pennsylvania in 13 and 27 days; wood thrushes returned to Pennsylvania from Honduras or Nicaragua in 13–29 days (n = 6; Stutchbury et al. 2009); similarily, veeries travelled from South America to Delaware in 17–33 days (n = 5; Heckscher et al. 2011). More data will most probably become available very soon.

In the recent years, the season of spring migratory arrival in most passerines, long-distance migrants included, has shifted towards earlier dates (Moritz 1993; Mason 1995; Sokolov et al. 1998; Sokolov and Payevsky 1998; Sokolov 2000, 2006; Crick and Sparks 2006; Gordo 2007; Møller et al. 2010). In western Europe, where winters have become much milder and spring phenology has advanced a lot, the advancement of arrival dates of long-distance migrants is insufficient, resulting in the mismatch between time of reproduction and the peak of resources: even though birds breed earlier in calendar dates, phenologically they breed later than in earlier years (Both and Visser 2001; Both et al. 2005, 2010). Many passerines, e.g. pied flycatchers in the Netherlands, breed now phenologically too late, in the suboptimal season. However, in other regions, where spring phenology has not (yet?) advanced so much as it did in western Europe (i.e. in the Urals or in Siberia), no such mismatch is currently observed (Sokolov and Gordienko 2008; Ananin and Sokolov 2009).

8.6 Conclusions

1. Passerine migrants usually stop over for 1–15 days. Sometimes, especially before and just after crossing large ecological barriers (large water bodies, deserts) stopovers may be longer and reach 20–25 days. Significant proportions of migrants stop over for one day only and continue migration on the first night after arrival. When studying stopover behaviour by stochastic capture-mark-recapture models, it is most useful to estimate the proportion of 'transients' (migrants that make one-day stopovers) and the mean stopover duration of non-transients. However, it should be kept in mind that capture-mark-recapture models tend to overestimate the number of transients, whereas radio-tagging results in more realistic estimates.
2. The mean fuel deposition rate (FDR) during migratory stopovers varies between zero and even negative values to 0.5% of lean body mass per day, but theoretically under ideal conditions may be as high as 10% per day. FDR does not remain constant during the stopover: it is low or even negative during the first 1–2 days upon arrival, then increases and may decrease again in the last

days of a prolonged stopover, especially if migratory departure is delayed by adverse weather. The mean FDR is often inversely related to the initial body mass. During the autumn migratory season FDR may increase with the progress of the season, even when food abundance and availability decline. FDR broadly varies on the individual basis; a large proportion of its variation is not explained by the extrinsic factors.

3. The relationships between the main energetic parameters of migratory stopovers (fuel deposition rate, stopover duration and departure fuel load) are usually described in the framework of the optimal migration theory. This theory is the accepted paradigm in the bird migration research, and the study of movement ecology and behaviour and of the evolution of migration is usually performed in this framework. Analysis of the original and literature data calls the correctness of this framework in question. The idea of the U-shaped relationship between flight speed and flight power, which is fundamental for the optimal migration theory, is not supported by the empirical data. In long-distance songbird migrants, i.e. birds adapted to endurance flapping flight, energy cost of flight is independent of its speed in a broad range of flight velocities. Up to the fuel loads of 25–30%, the transport of extra load (fuel) is nearly free, and the potential flight range is directly proportional to fuel load. The existing concept is a result of application of the fixed-wing flight theory to the avian flight which is non-stationary.

   A serious weakness of the optimal migration theory is that its predictions are difficult to test. By varying the initial assumptions, it is possible to change model predictions in such a way that they will agree to practically any field data. However, the idea of finding qualitative and quantitative relationships between the energetic parameters of migratory stopovers is a very useful one; it has greatly advanced avian migration research.
4. The correct selection of the optimal habitat during migratory stopovers plays a great (often decisive) role for safe and successful migration. Nocturnal migrants usually solve this problem by using visual cues when landing after migratory flights. The number of individuals that fail to recognize correctly their habitat is usually small; it usually happens under the conditions of poor visibility (rain, fog). Wetland passerines use for habitat recognition not only visual but also acoustic stimuli. They respond not only to the conspecific vocalisations but also to heterospecific song typical of the certain habitat (i.e. to the acoustic habitat markers). The response to song is either inherited or is developed early in life. Migrating passerines often have to utilise habitats significantly different from those they prefer during breeding. Apart from the immediate habitats, the broader landscape context is also an important factor which shapes the quality of a stopover site for migrants.
5. An extreme case of selection of an optimal habitat patch in the inhospitable matrix is stopping over on islands and in oases. The hypothesis that oases are ecological traps for stopover migrants that cannot refuel there because of their small carrying capacity and competition is not supported by the field data. Even

on small islands and in small oases the mean FDR is usually not much lower than in continuous habitat.

6. Daytime movements of nocturnal migrants do not refer to migratory activity and are habitat- and foraging-related. Usually they occur towards optimal habitats and are not generally directed towards the goal of migration. The exception is to so-called morning flights that at least in some cases are performed to compensate for wind drift during the preceding long-distance migratory flight, and migratory movements of migrants with the mixed rhythm of diel activity (*Turdus* thrushes, goldcrests, bramblings *Fringilla montifringilla* etc.).

   Spatial behaviour of passerines at migratory stopovers is very variable. Some species, e.g. the European robin, after ceasing migratory flight move across hundreds of metres and either resume flight on the first or second night upon arrival or occupy a restricted home range (some species defend it) and remain there until the end of stopover. Other species, e.g. pied flycatchers in spring, move broadly looking for locally abundant food throughout the stopover period. It is not inconceivable that the same species of migrants may employ different spatial strategies depending on the ecological conditions at stopover. The main factor that governs spatial stopover behaviour is the spatio-temporal distribution of the preferred food. If the food is relatively uniformly distributed in space and predictable in time, the migrants tend to occupy (and sometimes to defend) small home ranges. Species that utilise patchily distributed and unpredictable food sources make broad movements.
7. The current concept of the temporal schedule of nocturnal migratory flights is not accurate. The synchronised departures at the beginning of the night are only typical of migration during short nights (mainly in spring at temperate and high latitudes). When nights are long, many take-offs occur long after the end of the evening twilight. Fuel loads of departing migrants vary broadly and often are not significantly higher than the mean fuel loads of migrants at stopover. Some birds initiate nocturnal flights with rather small fuel loads. They may be making short flights in the migratory direction, and may be performing landscape-scale nocturnal flights aimed at habitat optimisation in reverse or any other direction.
8. The rate of energy expenditure in migratory flight in long-distance passerine migrants, adapted to endurance flights, is 6–7 times higher than their basal metabolic rate (BMR). This is much lower than the value of 10–12 BMR hitherto assumed (Dolnik 1995; Berthold 1996). The data from both free-flying birds and wind tunnel experiments provide evidence that migratory flights are energetically cheaper than it is usually believed.
9. The departure decision is made by a migrant under the influence of many factors that may be lumped into four main groups. These groups of factors may be regarded as relatively independent and represented as orthogonal axes in the four-dimensional hyperspace: (1) current energy status; (2) habitat quality expressed as FDR; (3) weather conditions, mainly wind assistance; and (4) position on the migratory route with respect to the individual schedule of migration. In this hyperspace there exists a four-dimensional region where take-

offs are permitted. The more favourable the weather is the smaller fuel load is sufficient for departure. The more is a migrant delayed the more it is motivated to migrate, even in poor weather conditions and with low fuel stores.

8.7 Perspectives of Research of Stopover Ecology and Behaviour of Passerines

The proposed model of stopover behaviour (four-dimensional hyperspace) has to remain qualitative at the current stage of bird migration research. One of the main problems that hinders making this model a quantitative one is the problem of habitat quality quantification: it is only possible to express it through the fuel deposition rate for the individuals that remained, but not for the ones that left. Equally difficult is it to estimate the position of an individual on the migratory route with respect to its unknown individual migratory schedule, which defines the urge to migrate (the ‘pure’ motivation, free from the influence of habitat quality, fuel stores and weather conditions). Currently even the qualitative model seems to be a step forward in our understanding of the principles of organisation of migration in passerines.

The studies of stopover ecology and behaviour started in the late 1980s and intensively developed in the following two decades. Great progress was achieved in estimating stopover duration by capture-mark-recapture statistics and especially by radio-tracking of small passerines. Capture-mark-recapture modelling made it possible to estimate statistically correctly the duration of stopovers on the basis of biologically realistic assumptions. It should be emphasized that this method permits not only estimation of the *mean* stopover duration but also to study the structure of its variation which is most important for the understanding how flights and stopovers alternate.

Radio-tracking studies made it possible to obtain unbiased estimates of stopover duration and of spatial behaviour of migrants at stopovers. However, the problem of obtaining unbiased estimates of fuel deposition rate remains elusive. To understand the principles of organisation of stopovers and flights (i.e. organisation of avian migration) it is not sufficient to have unbiased estimates of the mean FDR across the stopover period, which is by itself very difficult. It is necessary to have daily estimates of FDR for each day of stopover *without* multiple captures and *without* supplementary feeding, which bias the estimates so much that make them useless. This aim remains to be achieved.

Further progress in this research will most probably be achieved by the transition from the ground-based to satellite telemetry of small songbirds. Such data might allow us to know exactly the beginning and the end of migratory flights of individuals with known body mass and energy stores. This opportunity may become available in the foreseeable future, if the ICARUS initiative, aimed at satellite tracking of small animals, including birds, with conventional small VHF transmitters, is realised (Wikelski et al. 2007; Robinson et al. 2010). Another avenue of research which is already open is using light–dark loggers, the so-called geolocators which make it possible to track the movements of small birds, including passerines, with the accuracy of 100–200 km (Stutchbury et al. 2009, 2011; Bächler et al. 2010; Heckscher et al. 2011; Ryder et al. 2011; Bairlein et al. 2012). Results of geolocator studies that have been published since several years have already brought a wealth of data on migratory speed, number and duration of stopovers, geographic aspects of migratory movements, and are beyond doubt a major factor of progress in avian migration research. Another serious advancement would be a possibility to weigh the free-living birds without capture and without supplementary feeding. It would permit to measure FDR and its variation with habitat, landscape, current energetic condition, weather etc.

References

Ananin AA, Sokolov LV (2009) Long-term arrival trends of 54 avian species to Barguzinsky nature reserve in the northeastern Baikal area. Avian Ecol Behav 15:33–48

Arizaga J, Barba E, Belda EJ (2011) Effect of fuel load, date, rain and wind on departure decisions of a migratory passerine. J Ornithol 152:991–999

Bächler E, Hahn S, Schaub M, Arlettaz R, Jenni L, Fox JM, Afanasyev V, Liechti F (2010) Year-round tracking of small trans-Saharan migrants using light-level geolocators. PLoS ONE 5:e9566

Bairlein F (1985) Body weights and fat deposition of Palaearctic passerine migrants in the central Sahara. Oecologia 66:141–146

Bairlein F (1988) How do migratory songbirds cross the Sahara? Trends Ecol Evol 3:191–194

Bairlein F, Norris DR, Nagel R, Bulte M, Voigt CC, Fox JW, Hussel DJT, Schmaljohann H (2012) Cross-hemisphere migration of a 25 g songbird. Biol Lett 8: in press. doi:10.1098/rsbl.2011.1223

Battley PF (2006) Consistent annual schedules in a migratory shorebird. Biol Lett 2:517–520

Beck W, Wiltschko W (1988) Magnetic factors control the migratory direction of pied flycatchers (*Ficedula hypoleuca* Pallas). In: Ouellet H (ed) Acta XIX Congr Int Ornithol, vol 2. University of Ottawa Press, Ottawa

Begon M, Townsend CR, Harper JL (2006) Ecology: from individuals to ecosystems, 4th edn. Blackwell, Oxford

Bensch S, Nielsen B (1999) Autumn migration speed of juvenile reed and sedge warblers in relation to date and fat loads. Condor 101:153–156

Berthold P (1990) Spatiotemporal programmes and genetics of orientation. Experientia 46:363–371

Berthold P (1996) Control of bird migration. Chapman and Hall, London

Berthold P (2001) Bird migration: a general survey, 2nd edn. Oxford University Press, Oxford

Biebach H (1985) Sahara stopover in migratory flycatchers: fat and food affect the time program. Experientia 41:695–697

Biebach H, Friedrich W, Heine G (1986) Interaction of body mass, fat, foraging and stopover period in trans-Sahara migrating passerine birds. Oecologia 69:370–379

Blyumental TI, Gavrilov VM, Dolnik VR (1967) O prichinakh volnoobraznosti migratsii zyablika (On the causes of wave-like pattern of the chaffinch migration). Comm Baltic Comm Study Bird Migr 4:69–80

Bolshakov CV (1981) Rekonstruktsiya polnoy kartiny nochnogo prolyota i effektivnost' obnaruzheniya ego raznymi metodami (Reconstruction of the total picture of nocturnal passage and effectiveness of several methods of its estimation). Proc Zool Inst 104:95–123

Bolshakov CV (1992) Vechernie peremescheniya i start nochnogo migratsionnogo poleta u drozda-ryabinnika (*Turdus pilaris*): predvaritelnye rezultaty (Evening movements and the start of the nocturnal migratory flight in the fieldfare (*turdus pilaris*): preliminary results). Proc Zool Inst 247:18–42

Bolshakov CV (2002) The Palaearctic-African bird migration system: the role of desert and highland barrier of Western Asia. Ardea 90:515–523

Bolshakov CV (2003) Nocturnal migration of passerines in the desert-highland zone of Western Central Asia: selected aspects. In: Berthold P, Gwinner E, Sonnenschein E (eds) Avian migration. Springer, Berlin

Bolshakov CV, Bulyuk VN (1999) Time of nocturnal flight initiation (take-off activity) in the European Robin *Erithacus rubecula* during spring migration: direct observations between sunset and sunrise. Avian Ecol Behav 2:51–74

Bolshakov CV, Rezvyi SP (1998) Time of nocturnal flight initiation (take-off activity) in the European Robin *Erithacus rubecula* during spring migration: visual observations between sunset and darkness. Avian Ecol Behav 1:37–49

Bolshakov CV, Švažas S, Žalakevičius M (2002) Nocturnal migration of thrushes in the Eastern Baltic region. Akstis, Vilnius

Bolshakov C, Bulyuk V, Chernetsov N (2003a) Spring nocturnal migration of Reed Warblers *Acrocephalus scirpaceus*: departure, landing and body condition. Ibis 145:106–112

Bolshakov CV, Bulyuk VN, Mukhin A, Chernetsov N (2003b) Body mass and fat reserves of Sedge Warblers during vernal nocturnal migration: departure versus arrival. J Field Ornithol 74:81–89

Bolshakov CV, Chernetsov N, Mukhin A, Bulyuk VN, Kosarev V, Ktitorov P, Leoke D, Tsvey A (2007) Time of nocturnal departures in European robins, *Erithacus rubecula*, in relation to celestial cues, season, stopover duration and fat stores. Anim Behav 74:855–865

Both C, Visser ME (2001) Adjustment to climate change is constrained by arrival date in a long-distance migrant bird. Nature 411:296–298

Both C, Bijlsma RG, Visser ME (2005) Climatic effects on spring migration and breeding in a long-distance migrant, the pied flycatcher *Ficedula hypoleuca*. J Avian Biol 36:368–373

Both C, Van Turnhout CAM, Bijlsma RG, Siepel H, Van Strien AJ, Foppen RPB (2010) Avian population consequences of climate change are most severe for long-distance migrants in seasonal habitats. Proc R Soc London B 277:1259–1266

Bulyuk VN (2006) At what time of the day do passerine nocturnal migrants arrive at their breeding sites? Ardea 94:132–139

Bulyuk VN (2010) Autumn departure of adult nocturnally migrating passerines from their breeding sites. Avian Ecol Behav 17:3–12

Bulyuk VN, Mukhin A (2010) Do weather and lunar cycle influence the decision of juvenile reed warblers *Acrocephalus scirpaceus* to perform post-fledging nocturnal flights over the natal area? Acta Ornithol 45:27–32

Bulyuk VN, Tsvey A (2006) Timing of nocturnal autumn migratory departures in juvenile European robins (*Erithacus rubecula*) and endogenous and external factors. J Ornithol 147:298–309

Chernetsov NS (1999) Migratsionnye strategii kamyshevok *Acrocephalus* spp. v predelakh Evropy (Migratory strategies of *Acrocephalus* warblers within Europe). Dissertation, Zoological Institute RAS

Chernetsov N (2002) Efficiency of migratory stopovers of Song Thrushes *Turdus philomelos* and Redwings *T. iliacus* and their migration strategies in the Eastern Baltic. Avian Ecol Behav 9:15–22

Chernetsov N, Manukyan A (1999) Feeding strategy of Reed Warblers *Acrocephalus scirpaceus* on migration. Avian Ecol Behav 3:59–68

Chernetsov N, Manukyan A (2000) Foraging strategy of the Sedge Warbler (*Acrocephalus schoenobaenus*) on migration. Vogelwarte 40:189–197
Chernetsov N, Mukhin A (2001) Possible endogenous basis of juvenile dispersal in the reed warbler (*Acrocephalus scirpaceus*). Ring 23:173–177
Chernetsov N, Titov N (2001) Migratory stopovers of juvenile Blackcaps *Sylvia atricapilla* in autumn: stopover length, fuel deposition rate, and an attempt to predict departure body mass. Avian Ecol Behav 6:27–28
Chernetsov N, Bulyuk VN, Ktitorov P (2007) Migratory stopovers of passerines in an oasis at the crossroad of the African and Indian flyways. Ring Migr 23:243–251
Chernetsov N, Kishkinev D, Gashkov S, Kosarev V, Bolshakov CV (2008) Migratory programme of juvenile pied flycatchers, *Ficedula hypoleuca*, from Siberia implies a detour around Central Asia. Anim Behav 75:539–545
Conklin JR, Battley PF, Potter MA, Fox JW (2010) Breeding latitude drives individual schedules in a trans-hemispheric migrant bird. Nat Comm 1(67):1–6
Crick HQP, Sparks TH (2006) Changes in the phenology of breeding and migration in relation to global climate change. Acta Zool Sinica 52:154–157
Dolnik VR (1968) Rol' zhirovyh depo v regulyatsii metabolizma i povedeniya ptits vo vremya migratsii (Role of the fat depots in the regulation of metabolism and behaviour of the birds during migration). Zool Zhurnal 47:273–291
Dolnik VR (1975) Migratsionnoye sostoyanie ptits (Migratory disposition of birds). Nauka, Moscow
Dolnik VR (1990) Bird migration across arid and mountainous regions of Middle Asia and Kasakhstan. In: Gwinner E (ed) Bird migration. Springer, Berlin
Dolnik VR (1995) Resursy energii i vremeni u ptits v prirode (Energy and time resources in free-living birds). Nauka, St Petersburg
Dolnik VR, Blyumental TI (1967) Autumnal premigratory and migratory periods in the Chaffinch (*Fringilla coelebs coelebs*) and some other temperate-zone passerine birds. Condor 69:435–468
Ellegren H (1990) Autumn migration speed in Scandinavian bluethroats, *Luscinia s. svecica*, during autumn migration in Sweden. Ring Migr 11:121–131
Ellegren H (1993) Speed of migration and migratory flight lengths of passerine birds ringed during autumn migration in Sweden. Ornis Scand 24:220–228
Erni B, Liechti F, Bruderer B (2002) Stopover strategies in passerine bird migration: a simulation study. J Theor Biol 219:479–493
Forstmeier W (2002) Benefits of early arrival at breeding grounds vary between males. J Anim Ecol 71:1–9
Fransson T (1995) Timing and speed of migration in North and West European populations of *Sylvia* warblers. J Avian Biol 26:39–48
Fransson T, Jakobsson S, Johansson P, Kullberg C, Lind J, Vallin A (2001) Magnetic cues trigger extensive refuelling. Nature 414:35–36
Fusani L, Cardinale M, Carere C, Goymann W (2009) Stopover decision during migration: physiological conditions predict nocturnal restlessness in wild passerines. Biol Lett 5:302–305
Gavrilov VM (1981) Sutochnye izmeneniya metabolizma pokoya u ptits (Daily changes of the resting metabolism in birds). Ornithologia (Moscow) 16:42–50
Gavrilov VM (1985) Seasonal and circadian changes of thermoregulation in passerine and non-passerine birds: which is more important? In: Iljichev VD, Gavrilov VM (eds) Acta XVIII Congressus International Ornithol, Nauka, Moscow
Gavrliov VM (1980) Napravleniya bioenergeticheskikh adaptatsiy k sezonnosti klimata u ptits (Directions of the bioenegertic adaptation to climate seasonality in birds). In: Neufeldt IA (ed) Ekologiya, geografiya i okhrana ptits (Ecology, geography and conservation of birds). Zoological Institute, Leningrad
Gordo O (2007) Why are bird migration dates shifting? a review of weather and climate effects on avian migratory phenology. Clim Res 35:37–58

Grinkevich V, Chernetsov N, Mukhin A (2009) Juvenile Reed Warblers *Acrocephalus scirpaceus* see the world but settle close to home. Avian Ecol Behav 16:3–10

Gwinner E, Wiltschko W (1978) Endogenously controlled changes in migratory direction of the garden warbler, *Sylvia borin*. J Comp Physiol A 125:267–273

Gwinner E, Schwabl H, Schwabl-Benzinger I (1988) Effects of food-deprivation on migratory restlessness and diurnal activity in the garden warbler *Sylvia borin*. Oecologia 77:321–326

Hall-Karlsson KSS, Fransson T (2008) How far do birds fly during one migratory flight stage? Ring Migr 24:95–100

Hansson M, Pettersson J (1989) Competition and fat deposition in Goldcrests (*Regulus regulus*) at a migration stopover site. Vogelwarte 35:21–31

Heckscher CM, Taylor SM, Fox JW, Afanasyev V (2011) Veery (*Catharus fuscescens*) wintering locations, migratory connectivity, and a revision of its winter ranges using geolocators technology. Auk 128:531–542

Hedenström A, Pettersson J (1987) Migration routes and wintering areas of willow warblers *Phylloscopus trochilus* (L.) ringed in Fennoscandia. Ornis Fennica 64:137–143

Heinroth O, Heinroth K (1941) Das Heinfinde-Vermögen der Brieftauben. J Ornithol 89:213–256

Henshaw I, Fransson T, Jakobsson S, Lind J, Vallin A, Kullberg C (2008) Food intake and fuel deposition in a migratory bird is affected by multiple as well as single-step changes in the magnetic field. J Exp Biol 211:649–653

Henshaw I, Fransson T, Jakobsson S, Jenni-Eiermann S, Kullberg C (2009) Information from the geomagnetic field triggers a reduced adrenocortical response in a migratory bird. J Exp Biol 212:2902–2907

Jenni L, Schaub M (2003) Behavioural and physiological reactions to environmental variables in bird migration: a review. In: Berthold P, Gwinner E, Sonnenschein E (eds) Avian migration. Springer, Berlin

Jones P (1995) Migration strategies of Palaearctic passerines in Africa. Isr J Zool 41:393–406

Karlsson H, Nilsson C, Bäckman J, Alerstam T (2011) Nocturnal passerine migration without tailwind assistance. Ibis 153:485–493

Klaassen M, Kvist A, Lindström Å (2000) Flight costs and fuel composition of a bird migrating in a wind tunnel. Condor 102:444–451

Kokko H (1999) Competition for early arrival in migratory birds. J Anim Ecol 68:940–950

Koleček J (2006) Nadměrna hmotnost a tučnost u rakosnika velkeho (*Acrocephalus arundinaceus*). Sylvia 42:126–129

Kosarev V, Kobylkov D (2010) Razvitie predmigratsionnogo ozhireniya na mestakh razmnozheniya u trostnikovykh kamyshevok (*Acrocephalus scirpaceus*) pered nachalom osenney migratsii po dannym radiotelemetrii (Development of pre-migratory fattening at the breeding sites in Eurasian reed warblers (*Acrocephalus scirpaceus*) before the onset of autumn migration from the radio telemetry data). In: Kurochkin EN, Davygora AV (eds) Ornitologiya v Severnoy Evrasii (Ornithology in Northern Eurasia). Orenburg State University Press, Orenburg

Ktitorov P, Tsvey A, Mukhin A (2010) The good and the bad stopover: behaviours of migrant reed warblers at two contrasting sites. Behav Ecol Sociobiol 65:1135–1143

Kullberg C, Henshaw I, Jakobsson S, Johansson P, Fransson T (2007) Fuelling decisions in migratory birds: geomagnetic cues override the seasonal effect. Proc R Soc London B 274:2145–2151

Lindström Å (2003) Fuel deposition rates in migrating birds: causes, constraints and consequences. In: Berthold P, Gwinner E, Sonnenschein E (eds) Avian migration. Springer, Berlin

Lourenço PM, Kentie R, Schroeder J, Groen NM, Hooijmeijer JCEW, Piersma T (2011) Repeatable timing of northward departure, arrival and breeding in black-tailed godwits *Limosa l. limosa*, but no domino effects. J Ornithol 152:1023–1032

Mädlow W (1997) Durchzug und Rastverhalten des Rotkehlchens (*Erithacus rubecula*) im Herbst 1995 auf der Greifswalder Oie: Situation während eines Masseneinzugs. Seevögel 18:75–81

Markovets M, Yosef R (2005) Phenology, duration and site fidelity of wintering bluethroat (*Luscinia svecica*) at Eilat, Israel. J Arid Environ 61:93–100

Mason CF (1995) Long-term trends in the arrival dates of spring migrants. Bird Study 42:182–189

Meijer T, Mohring FJ, Trillmich F (1994) Annual and daily variation in body mass and fat of starlings *Sturnus vulgaris*. J Avian Biol 25:98–104

Merkel FW (1938) Zur Physiologie der Zugunruhe bei Vögeln. Ber Vereins Schles Ornithol 25:1–72

Merkel FW (1958) Untersuchungen zur künstlichen Beeinflussung der Aktivität gekäfigter Zugvögel. Vogelwarte 19:173–185

Mills AM, Thurber BJ, Mackenzie SA, Taylor PD (2011) Passerines use nocturnal flight for landscape-scale movements during migratory stopover. Condor 113:597–607

Møller AP, Fiedler W, Berthold P (eds) (2010) Effects of climate change on birds. Oxford University Press, NY

Moreau RE (1961) Problems of Mediterranean-Saharan migration. Ibis 103a:373–427

Moritz D (1993) Long-term monitoring of Palaearctic-African migrants at Helgoland (German Bight, North Sea). In: Wilson ET (ed) Proceedings of 8th Pan-African Ornithological Congress. Musée Royal de Afrique Centrale, Tervuren

Mukhin A (2004) Night movements of young reed warblers (*Acrocephalus scirpaceus*) in summer: is it postfledging dispersal? Auk 121:203–209

Mukhin A, Kosarev V, Ktitorov P (2005) Nocturnal life of young songbirds well before migration. Proc R Soc London B 272:1535–1539

Newton I (2008) The migration ecology of birds. Academic Press, London

Ottosson U, Waldenström J, Hjort C, McGregor R (2005) Garden Warbler *Sylvia borin* migration in sub-Saharan West Africa: phenology and body mass changes. Ibis 147:150–157

Panov IN (2011) Overlap between moult and autumn migration in passerines in northern taiga zone of Eastern Fennoscandia. Avian Ecol Behav 19:33–64

Panov IN (2012) Migratsionnye strategii vorobyinykh ptits v severnoy tayge Vostochnoy Fennoskandii (Migratory strategies of passerines in the northern taiga of Eastern Fennoscandia). Dissertation, Institute of Ecology and Evolution RAS

Paradis E, Baillie SR, Sutherland WJ, Gregory RD (1998) Patterns of natal and breeding dispersal in birds. J Anim Ecol 67:518–536

Pettersson J (1983) Rödhakens *Erithacus rubecula* höstflyttning vid Ottenby. Vår Fågelvärld 42:333–342

Raess M (2008) Continental efforts: migration speed in spring and autumn in an inner-Asian migrant. J Avian Biol 39:13–18

Richardson W (1978) Timing and amount of bird migration in relation to weather: a review. Oikos 30:224–272

Richardson W (1990) Timing of bird migration in relation to weather: updated review. In: Gwinner E (ed) Bird migration. Springer, Berlin

Robinson WD, Bowlin MS, Bisson I, Shamoun-Baranes J, Thorup K, Diehl R, Kunz TH, Mabey S, Winkler DW (2010) Integrating concepts and technologies to advance the study of bird migration. Front Ecol Environ 8:354–361

Ryder TB, Fox JW, Marra PP (2011) Estimating migratory connectivity of gray catbirds (*Dumetella carolinensis*) using geolocators and mark-recapture data. Auk 128:448–453

Salewski V, Bairlein F, Leisler B (2002) Different wintering strategies of two Palearctic migrants in West Africa—a consequence of foraging strategies? Ibis 144:85–93

Salewski V, Schmaljohann H, Liechti F (2010) Spring passerine migrants stopping over in the Sahara are not fall-outs. J Ornithol 151:371–378

Schaub M, Jenni L (2000) Fuel deposition of three passerine bird species along migration route. Oecologia 122:306–317

Schaub M, Liechti F, Jenni L (2004) Departure of migrating European robins, *Erithacus rubecula*, from a stopover site in relation to wind and rain. Anim Behav 67:229–237

Schmaljohann H, Naef-Daenzer B (2011) Body condition and wind support initiate the shift of migratory direction and timing of nocturnal departure in a songbird. J Anim Ecol 80:1115–1122

Schmaljohann H, Liechti F, Bruderer B (2007a) Songbird migration across the Sahara: the non-stop hypothesis rejected! Proc R Soc London B 274:735–739

Schmaljohann H, Liechti F, Bruderer B (2007b) An addendum to 'songbird migration across the Sahara: the non-stop hypothesis rejected!'. Proc R Soc London B 274:1919–1920

Schmaljohann H, Becker PJJ, Karaardic H, Liechti F, Neaf-Daenzer B, Grande C (2011) Nocturnal exploratory flights, departure time, and direction in a migratory songbird. J Ornithol 152:439–452

Sokolov LV (1997) Philopatry of migratory birds. In: Turpaev TM (ed) Physiology and general biology reviews, vol 11. Harwood Academic Press, Amsterdam

Sokolov LV (2000) Spring ambient temperature as an important factor controlling timing of arrival, breeding, post-fledging dispersal and breeding success of Pied Flycatchers *Ficedula hypoleuca* in Eastern Baltic. Avian Ecol Behav 5:79–104

Sokolov LV (2006) Effect of global warming on the timing of migration and breeding of passerine birds in the 20th century. Entomol Rev 86:S59–S81

Sokolov LV, Gordienko NS (2008) Has recent climate warming affected the dates of bird arrival to the Il'men reserve in the southern Urals? Russ J Ecol 39:56–62

Sokolov LV, Kosarev VV (2003) Relationship between timing of arrival of passerines to the courish spit and North Atlantic oscillation index (NAOI) and precipitation in Africa. Proc Zool Inst 299:141–154

Sokolov LV, Payevsky VA (1998) Spring temperatures influence on year-to-year variations in the breeding phenology of passerines on the Courish Spit, Eastern Baltic. Avian Ecol Behav 1:22–36

Sokolov LV, Markovets MY, Shapoval AP, Morozov YG (1998) Long-term trends in the timing of spring migration of passerines on the Courish Spit of the Baltic Sea. Avian Ecol Behav 1:1–21

Stutchbury BJM, Tarof SA, Done T, Gow E, Kramer PM, Tautin J, Fox JW, Afanasyev V (2009) Tracking long-distance songbird migration by using geolocators. Science 323:896

Stutchbury BJM, Gow E, Done T, MacPherson M, Fox JW, Afanasyev V (2011) Effects of post-breeding moult and energetic condition on timing of songbird migration into the tropics. Proc R Soc London B 278:131–137

Szulc-Olech B (1965) The resting period of migrant robins on autumn passage. Bird Study 12:1–7

Terrill SB (1990) Ecophysiological aspects of movements by migrants in the wintering quarters. In: Gwinner E (ed) Bird migration. Springer, Berlin

Titov N (1999) Individual home ranges of Robins *Erithacus rubecula* at stopovers during autumn migration. Vogelwelt 120:237–242

Tøttrup AP, Klaassen RHG, Strandberg R, Thorup K, Kristensen MW, Jørgensen PS, Fox J, Afanasyev V, Rahbek C, Alerstam T (2012) The annual cycle of a trans-equatorial Eurasian-African passerine migrant: different spatio-temporal strategies for autumn and spring migration. Proc R Soc London B 279:1008–1016

Tsvey A, Bulyuk VN, Kosarev V (2007) Influence of energy condition and weather on departures of first-year European robins, *Erithacus rubecula*, from an autumn migratory stopover site. Behav Ecol Sociobiol 61:1665–1674

Vega Rivera JH, Rappole JH, McShea WJ, Haas CA (1998) Wood thrush postfledging movements and habitat use in northern Virginia. Condor 100:69–78

Vega Rivera JH, McShea WJ, Rappole JH (2003) Comparison of breeding and post breeding movements and habitat requirements for the scarlet tanager (*Piranga olivacea*) in Virginia. Auk 120:632–644

Wang Y, Moore FR (1993) Relation between migratory activity and energetic condition among thrushes (turdinae) following passage across the Gulf of Mexico. Condor 95:934–943

Wang Y, Chang JC, Moore FR, Su LY, Cui LM, Yang XF (2006) Stopover ecology of red-flanked bush robin at maoershan of Northeast China. Acta Ecol Sin 26:638–646

Wikelski M, Kays RW, Kasdin NJ, Thorup K, Smith JA, Swenson GW Jr (2007) Going wild: what a global small-animal tracking system could do for experimental biologists. J Exp Biol 210:181–186

Wiltschko W, Wiltschko R (1992) Migratory orientation: magnetic compass orientation of garden warblers (*Sylvia borin*) after a simulated crossing of the magnetic equator. Ethology 91:70–79

Yamaguchi N, Tokita K-I, Uematsu A, Kuno K, Saeki M, Hiraoka E, Uchida K, Hotta M, Nakagawa F, Takahashi M, Nakamura H, Higuchi H (2008) The large-scale detoured migration route and the shifting pattern of migration in Oriental honey-buzzards breeding in Japan. J Zool 276:54–62

Yamaguchi NM, Arizawa Y, Shimada Y, Higuchi H (2012) Real-time weather analysis reveals the adaptability of direct sea-crossing by raptors. J Ethol 30:1–10

Yohannes E, Biebach H, Nikolaus G, Pearson DJ (2009a) Migration speeds among eleven species of long-distance migrating passerines across Europe, the desert and Eastern Africa. J Avian Biol 40:126–134

Yohannes E, Biebach H, Nikolaus G, Pearson DJ (2009b) Passerine migration strategies and body mass variation along geographic sectors across East Africa, the Middle East and the Arabian Peninsula. J Ornithol 150:369–381

Index

N. Chernetsov, *Passerine Migration*, DOI: 10.1007/978-3-642-29020-6,